ARMY BASIC TRAINING

ARMY BASIC
TRAINING

Bryan J. Stanley

VANTAGE PRESS
New York

FIRST EDITION

All rights reserved, including the right of
reproduction in whole or in part in any form.

Copyright © 1993 by Bryan J. Stanley

Published by Vantage Press, Inc.
516 West 34th Street, New York, New York 10001

Manufactured in the United States of America
ISBN: 0-533-10281-2

Library of Congress Catalog Card No.: 92-93051

0 9 8 7 6 5 4 3 2 1

To all who have served, are serving, and will serve in the greatest defender of human freedom the world has ever seen: the U.S. Army

Preface

I wrote this book for several reasons. One was to inform people of what basic training is like. There are probably many people who would like to know what it is like: New recruits to the army will like knowing what Basic is like before they go. Former army personnel will have a chance to relive their bit, and that could be a bittersweet experience. People who have never been in the army may be interested in just what it is like on a day-to-day basis. Finally, military historians in the future may find the book of some interest.

I started the book in October of 1981. I did not go into the army until December of '81, but I had some interaction with the army before actually going in and I felt it would give a smoother style to my writing in the time sequence of actually entering the army.

It is now August of 1991, nearly ten years after I entered the army. You might be wondering why I waited so long to get to work on this. Well, the truth is that I lost the manuscript of my work and only recently discovered it. So let's get started.

ARMY BASIC TRAINING

14 October 1981

I am sitting in a jet on the ground at Eau Claire, Wisconsin, flight 531. The jet is just starting to move, hiss, and scream a bit, and in a short time we will be airborne and heading for Minneapolis. I was originally supposed to be heading to the city by bus, but at the last minute the army changed the plans and got me a complete flight to Grand Forks, North Dakota, for my flight physical. I am looking forward to partying in the city.

Wow! Here we go off the ground. This is great. There are fall colors out, but that does not help much, since the sky is overcast and I cannot see anything. Wow, we have just passed through the clouds and are above them. What a beautiful world up here, a sea of cotton and blue sky above.

We land in Minneapolis and, after leaving 531 and getting in the terminal, I find some hot dogs to eat in a small restaurant, expensive at $1 each.

The flight to Grand Forks is grand: clear skies, sun, thousands of beautiful farms and pastures. Many crops are being harvested. A brown and green quilt studded with rhinestones of silver farm buildings. Many black, grotesquely shaped lakes and serpentine rivers. Very indicative of glacial terrain. I can see a hundred miles. This is a rich blend of cultural and natural features. The farms appear much larger on the west side of Minnesota. Now we are getting lower and farmhouses no longer seem like rhinestones, just toys. There is no land like America.

I make it to Grand Forks and am feeling good. I had expected to walk to the air base from the commercial airport.

I was in for a shock. I wander around the pool hall–sized terminal for about fifteen minutes, trying to decide what to do. Then I see a blue air force car and decide to ask for a ride. Luckily, the driver says yes; he also says the base is about ten miles away.

Well, I made it here to the BOQ (bachelor officers' quarters) and was put in room 118-B. It's not a single, however; there are three airmen. Here is one now, but he will be moving out. All of these people here are air force and are somewhat shocked that I am army. Well, they all just left and are going out partying. Word is that the army can find all kinds of ways to screw you. I am property, they say, dog tags. I might be sent to Saudi Arabia, or something. Well, I hope not. But the only way to find out is to try it, I guess.

I ate at the Red River Inn here at Grand River Air Force Base, a name for the barracks mess hall. Much different than I imagined. It looked like a civilian cafeteria, not the Gomer Pyle style I thought I would see. Good food and a modern, comfortable setting. After eating I took a stroll around the base a bit. The base is military, but not like I imagined. People are alive and they seem contented. I found a First National Bank, a credit union, a movie theater, a recreation center, and a couple of shopping areas. I went back to my room and began to reflect on the reasons that I was joining the army and came up with several. I guess one reason is to buy some time to realize what I really want to do with my life. Two, to travel a bit. Three, live in Europe. Four, learn a foreign language. Five, save some money. Six, get some educational benefits. Seven, learn what the service is like. And eight, do something different.

15 October 1981

My physical exam at Grand Forks Air Force Base was not

so great. An EKG is a rather strange test. The doctor attached twelve sensors to my legs, arms, and chest and then listened with a machine for two minutes. This test is supposed to get a picture, of sorts, of my heart and its activity. I also got my hearing checked again, weight, height, blood, eyes, blood pressure.

I thought I was doing well until I finally met the flight surgeon. He queried me about my hay fever. How long I had had it and so on. It seems he didn't think an air traffic controller should have that and strongly urged my disqualification in that area. Well, of course I was shocked. He said he didn't know if the army would DQ me, but if I was in the air force he would immediately.

Hmmmph, I'll wait and see what the army has to say about it when I get back to La Crosse. Maybe I will have to opt for something different. Maybe I'll go for the foreign language thing, but I would need to go in for four years. This is being written at Minneapolis International Airport.

16 October 1981

To conclude my travel from Minneapolis, it was the most beautiful 130 miles I have ever traveled. By Republic Air to La Crosse south, gliding above the mighty Mississippi River. Skies were perfectly clear and the oaks, hickories, elms, aspens, and shrubs were in their prime autumn foliage. Nature is quite the artist.

It is difficult to put down in words what I saw, although I know Mark Twain could do it. The Mississippi was wide all the way, quiet, dark, braided from horizon to horizon, paralleled by the colorful bluffs. I have never seen such beautiful farms. Green and gold pastoral squares were on these ancient prairies. Huge gouges of forested coulees surrounded the peninsular farms. I would say that about 50 percent of the land was cultivated and about 50 percent forested coulees.

This view was with us all the way to La Crosse. Little towns bedecked the valley along the way: Red Wing, Pepin, Alma, Winona, and Trempealeau. How clean they looked! How organized! One thing that probably stood out the most to me in this area was the many contrasts. Farm forest, some level land, some steep land, some twists and turns, some consistently regular, angular, parallel cultural shapes; dark, limpid, forests and green and brown fields and pastures; lush, marshy braided river valley versus dry forests and farmlands. All of it very, very beautiful. My enlistment in the army is already beginning to pay off.

My last day in La Crosse, Wisconsin, was November 30, and in between the time I came back and the day I left La Crosse for good I signed up as a 98 golf (98G) with the army. That's a job where you listen to foreign communications. I signed up to study German and would have to go to language school for about eight months, followed by training in Texas. Yes, I was definitely going to do some traveling.

I was disqualified as an air traffic controller because of my hay fever. I had to sign up for four years with the army. I was signed up for three as an ATC. I took the language test up in Minneapolis and thought it was really difficult. I did not think I passed and was really depressed, wondering what I was going to do now. I was broke, with no future, and thought my future rested with the army. Then a recruiter came in and said I had passed, and that pleased me very much.

The night before I left I got together with a few friends to party. Some had been in the service, and they told me their stories. One guy, Tom, talked about how the USMC knocked him around and how he came back to La Crosse once for thirty days and regretted it. He was glad he went in, however.

30 November 1981

I am sitting in the Winona, Minnesota, bus station full of

nauseating diesel fumes; my nose and head are full of them. Luckily, it is a short wait here—less than five minutes. I am leaving La Crosse, leaving the coulee region, leaving Wisconsin. I have done this leaving before, but it was easier then than now. I left for college in '75, I left for South Dakota in '78, and I left for Michigan in '79. Each time I am more aware of what I am leaving. I would not say it causes a pain, but rather a melancholy feeling. A profound respect and knowledgeable feeling of what I am leaving. I know that no matter where I go I won't find a place anything like this. Some might say I am simply showing homesickness and that everyone feels the same about their home.

That may be correct, but La Crosse is something special. With its clean air, clean streets, and clean water. With its urban forests, Trane Company, and Heileman brewery. With its immigration history, Germans and Norwegians. With its Oktoberfest and riverboats, the Mississippi River, and part of the driftless area where no glacier has been. And yet there is nothing for me here: no work, no job, nobody, I am sorry to say.

I think the whole idea of me ever settling in La Crosse is just a fantasy. Who knows, maybe I'll find some other nice place after living in the army for a while.

I wonder what is in store for me in the army. I wonder if I'll succeed or fail. A change? A change for sure. I wonder how much I'll change? Will I maintain my independence or forever lose that? Only time will tell, as it always does.

1 December 1981

I am writing in Army Liaison at Afees (recruit processing) in Minneapolis. I have some time to finish writing about my bus trip to Minneapolis. I learned a little about the army last night by eavesdropping on a conversation between the bus driver and a guy who was going into the army. The bus driver had been in the army in the early 1960s for five years. He said he liked the army; he was in Europe. Had been to Helsinki, Copenhagen, Stockholm, Oslo, Brussels, Hamburg, and Munich. He said he liked the trains over there.

When I got to Minneapolis, I called a former girlfriend and went out to eat. I stayed at her house last night. She dropped me at the hotel at 6:30 this morning. I had breakfast and was the only one in the room. The waitress fetched me coffee, eggs, juice, and water. She complained that I got the service of twenty recruits. I simply labeled her a bitch.

I walked through the snow to the Afees building. It was wet out, the temperature about thirty-three degrees. Had to avoid curbs and cars splashing slush. I reached the Afees building and dropped my bags in the army liaison office. I was ordered upstairs for a spot check. I waited with sixty other recruits. I waited an hour and was so bored, I tried to go downstairs to get a book, but a med tech sergeant named Frazier said no and shook his head. I noticed he slouched and had dark, short, greasy hair. He could be a consummate asshole.

I sat down and waited another half-hour. Finally, a doctor came in and told us to strip to our underwear. The doctor was old and hoary. He told us to go in the room and line up one block behind the yellow line. He checked our weight, among other things. I weighed 155 pounds. I have not gained a pound since high school in 1973.

We got dressed and went downstairs to wait some more. I thought about what a friend had said about the army: "Hurry up and wait, that's all we did."

There were seven army recruiters in the room. Each had his own office with no door, simply a partition. Listening to army recruiters, they sound congenial and friendly. They do not worry about the business end of the deal. They just deal with logistics of recruits.

I have been thinking about how that sergeant ordered me to sit down and wait. I'll admit I got a little angry at first. Now I think I must resolve to put all personal thoughts aside and remember not to take action on something without first being ordered since I am now in the army. It's a good idea for efficiency, but one should be conscious that one is obeying a particular order and that a command is not going to run or ruin one's life.

I think one must have some control of some area of one's own life where one takes the initiative. Either a diary or a small book or counting one's money and thinking about what one is going to do with it. Chain of command and obeying are necessary for order and efficiency. But the pecking order can be abused by both commander and commandee. A commander can be power-hungry and order everything and let it go to his head and lose his humility. A commandee can abuse his subservient role by not taking any creative action his whole life. By not managing his free time or not taking the time to plan his future. A commandee must maintain some of his creative identity, not by choosing which command to follow (which will lead to disaster), but by planning his personal budget and thinking about how this time in the service will benefit him in the future. This leads me to think of two quotes by Ben Franklin. One is: "He that cannot obey cannot command," and two, "To be humble to superiors is duty, to equals courtesy, and to inferiors nobleness."

I notice while waiting in this army liaison the variety of individuals, and I wonder how effective the army's equalizing program at Basic Training will be. Short haircut, white T-shirt, khaki coat, khaki pants, black boots, same bed, same D.I. (drill instructor), same everything. It makes me feel good already that I will not have to worry about inferiority. I will have a stripe, but that's about it. I will feel good just starting out the same. In the real world so often so many things work against a person such as his hair, if it is thin or gone or not an attractive color. The recruits here are about fifteen now; white windbreaker, blue jeans, red hair, blond hair, shaggy hair, neat hair, cool hair, and gross hair. Ski jackets, sport coats. One person looks classy, another scraggly and bummy. I am in between, dress shirt, dress jeans, and dress shoes. Yes, I think the equalizer program will work well.

Forms, many forms, many contracts, many signatures. Most of the time I don't know what I am signing. Photocopies and more contracts. At this point most are contracts. I have been warned for my own good to keep a copy of everything. I am and everybody else should, too. Yesterday I bought an expanding folder that ties shut. I just hope it is big enough.

Now there are few people remaining in this army room of human processing. The recruits who remain are in good spirits and maintain congenial conversations. There is the definite tension of a traveling atmosphere. Everyone here is ready to ship out today. Fort Dix, Fort Knox, Fort Rucker, Fort Leonard Wood, and others. There are at least forty suitcases and travel cases of different kinds. Hand-me-down duffel bags from World War II, new American Touristers, and sleek brown leather cases. Maybe going-away presents. What a rainbow of travel. One can only wonder where they have been, what miles they have traveled, and what battles they have fought. A marvelous bit of inspiration.

Everything about this room says army. The U.S. map of American bases, old and worn from years of use and service. Posters of Europe, Eiffel Tower, London changing of the guard, medieval bridges, castles, barges and villages along the Rhine. What young person can turn away from these attractions, these adventures, these chances? Never before have so many young people had so much to choose from. So much travel with just a little commitment. Who wouldn't trade the dirty pinball hangouts and blitzed out zodiac rooms full of electronic machines for these natural wonders? To see the world, to see new cultures . . . WOW!

It's 12:55 P.M. I've been waiting to be sworn in now for twenty-five minutes. They moved us in here from the liaison office to this small swearing-in room. There are nine of us. It seems a few are lost. Now it's 1:00 P.M. All of us are here now, and we solemnly swear to defend the U.S. Constitution and obey the president—nothing wrong with that, I guess.

Now we are sent out into the hall to wait for our call to transportation. They say it will get here at 1:15 P.M.

Well, they misjudged by only an hour and fifteen minutes, and at 2:30 P.M. they begin to call our groups and group leaders to come up and get our tickets. An attractive female soldier, about 5 feet 6 inches tall with short dark hair and a loud, reverberating voice, calls out, "Army, Fort Dix; Marines, San Diego; Navy, Great Lakes, . . . " and so on. This wait was very boring.

At this point the recruits look very average. A few have shaved heads already, no doubt to cheat the army barbers out of some sadistic pleasure. Others have mostly long hair and typical eighties dress: blue Levis. Many, many smokers. Most just sit hunched in their chairs with hands under chins and elbows on knees. When . . . can . . . we . . . go . . . ? Boring . . . boring . . . boring.

We finally got our tickets and got on a big school bus with a great stereo system, and off we went with a very loud murmur of voices and Beatles' music playing. It took us about twenty minutes to get to the airport, but we made it. After securing our tickets and baggage, we rested and waited three hours for our 5:45 flight to St. Louis. We arrived in the visiting servicemen's quarters at 3:30 P.M. We all had a snack there of hot dogs and chips. There was pop to drink and coffee, tea, and hot chocolate. There was a small library of *National Geographic* and other magazines and paperbacks. There was a game room upstairs and pool and electronic games. This service was free and was very welcome to us. It is nice to know that somebody cares about you when you are traveling and alone. There was also a TV to watch and nice comfy leather seats to sit on. There were many letters on the wall from past servicemen who've passed through here commending the volunteers who run this service. On the wall was a world map, and on it were hundreds of pins marking the locations of traveling servicemen.

After that we got on our jet for St. Louis. The plane was a bit crowded, with ninety-seven passengers, but I managed to talk my way out of getting a smoking-section seat. I am writing now on the plane, and we are taxiing down the runway, a smooth, undulating ride. We will be off in a moment and on our way for a one-hour trip. We are turning . . . We are accelerating . . . We are off!

It's still 1 December 1981. As flights go I still can't say I relish air travel. Nausea, plugged ears, cramped sitting, and headaches. But it is fast and the misery is probably less in the long run compared to other travel, except possibly ocean liners. We arrived safe from Minneapolis on flight 466, Northwest Orient. Now we have to catch a bus to Fort Leonard Wood in twenty minutes. Then the assaults begin, I bet. You are in the army now.

2 December 1981

We arrived at Fort Wood at 1:30 A.M. in a Greyhound bus. We were picked up by an army bus, which was that wonderful color OD (olive drab) green. We were served dinner right away; it was pork chops, mashed potatoes, lettuce, and cottage cheese. It wasn't bad, but it gave me a flatulent atmosphere.

All of my highest expectations of Leonard Wood have been filled and exceeded. It is fine. Vintage army. In a single word . . . drab. That is Fort Wood. Drab, that says it all. By all definitions of the word. This mighty Fort, Leonard Wood, brought humbly to its knees with one swift drab. I can see it now; fifty years from now while I am reclined in my easy chair, hoary with age, I will be able to squint back into the faded past and know that they redefined drab to an unprecedented degree. It might even become a cliché. I will read in a little corner of some daily newspaper a small story about some miners on some forgotten moon saying that the moon is "as drab as Fort Leonard Wood in December."

I still have my hair, as do most of us, but will lose it today. That depresses me. I think things could go on with longer hair.

They do not want anyone to have privacy here. I've figured that out already. They have no stalls for the toilets. This is the great secret of demoralizing the last vestige of a man's castle. The walls removed from the one great resting spot in an unrestful world. If they want to humble us they got the right idea. I can give the army credit for brains already.

It's 6:15 P.M. Very little free time to jot down notes. I am having trouble already remembering what happened all day. We filed out at 9:30 A.M. We listened to another platoon march by us. They were very good at marching and had very good cadence, very precise and very loud. We were marched over to wait for haircuts; we had choice of long or short. Long is 1/2 inch; short is a shave. Many recruits looked bald and funny. Most were good-natured about it. Everybody laughs at themselves and each other. After getting my haircut, I, along with the other recruits, had to wait two hours under mostly cloudy skies and small patches of blue and sun. We talked constantly and twice the sergeant (a mellow guy) told us to quiet down. While we were waiting, small groups of soldiers would walk by and razz us and take off their hats and rub their heads. One thing good about these long waits is that it gives us a chance to get to know each other.

For lunch today we had breaded veal Parmesan, carrots,

choice of cake: mince, coconut, or apple . . . very good. Parmesan very good. Must eat very fast, no talking, must eat with coats on. We toast each other with silent talk of eyes. Usually I take much food but don't enjoy it, even though it is good. Should probably take less and enjoy it more.

We were finally relieved and I went up to relax in bed. I passed out and in a short time was rousted, and we had to go to a meeting with a drill instructor. At this point we have not commenced Basic Training yet; we are still being processed. The sergeant answered questions about our gear and some aspects of military law. Also demands to be called "Sergeant," not "Sarge," not "hey, man," and not "Hey, bro." He is dressed in Viet camo and has a Smokey Bear hat, like a ranger. He has pronounced cheek-bones, is very stern, is clear in his speech, and his name is Skinner. What's he going to do, skin us alive? He makes note that *we* enlisted to become *soldiers*. If we don't want to, it will soon be apparent.

We go to dinner—roast beef, tender and tasty, mashed potatoes, peas, cottage cheese, Jell-O, chocolate cake, milk, and soda, all very good. But still must eat with coat on and no talking.

After dinner we walk to break area near barracks. It's a gravel area about thirty feet by fifty feet with four oak trees. It's the only place where we are allowed to smoke. We salute as "Retreat" and "Hail to the Colors" are played. It's getting dark; we tell a few jokes, laugh. Our whole platoon is here, people from all parts of the U.S. I know about three or four well. Dan, 98-golf from Minneapolis, and Al, 98-G from Cincinnati, and another guy in radiology from Dayton. We talk of weather, army, Patton, rank in the army, and joke a bit. One guy, a pfc, is a nerd and typical court holder. He knows everything about everything. He prides himself that he can tell us everything we are going to learn in basic. He is prior service. He tells how tough army is and what pussies marines are. Says he and four other army guys whipped a whole platoon of marines. We all wink at each other. Bullshitter is going to be a truck driver.

Break time over. Another meeting with a sergeant. A black man, he tells us to organize guard duty. Two guards will stand watch every hour on the half hour for an hour. Sounds like it will be fun. He tells us what to do. When we wake up on the quarter past the hour we get dressed. At twenty-five past we walk to building 2145 and sign in. We then return to barrack and relieve guards and stand watch for an hour. Anybody that knocks on door (now why would anyone do that out here?) is to halt and show ID. We are to ask for last four numbers of Social Security number, then ask why here and follow around until business is finished. If a maniac comes and enters building and refuses to leave, then guard should not force maniac to leave, because guard does not have that right. Guard should wake another recruit and send him to building 2145 to get MPs so they can remove maniac.

We return to barrack and we are assigned clean-up duties and guard duties. This discipline is necessary for efficiency, to be ready in an emergency. The deep issue is not that the floors are clean or grounds are clean or that the beds are made properly. The idea is that we can follow orders correctly and promptly. Our superiors must feel secure that they can depend on you to do your job. At some point a great deal of responsibility may fall on our shoulders, and many lives may depend on us.

We do not have much time; other men are cleaning latrines, mopping floors, and dusting. My squad will clean grounds tomorrow morning. Tonight lights out at 8:30 P.M. Tomorrow get up at 4:00 A.M. Personal hygiene between 4:00 and 4:45. Clean-up between 4:45 and 5:10. I wonder how we will be able to clean the grounds when it is dark outside. If we pass all inspections we may have a free weekend.

Basic Training is scheduled to start next Tuesday after our processing, getting uniforms, getting used to army, and getting in the groove. At this point we are sometimes referred to as bodies. Women, I noticed, are called ladies. Basic is eight weeks scheduled to break sometime in February. March 4 begins my language school in California. Well, I must dust now; bedtime is at 8:30 and I'm tired.

3 December 1981

It's 6:30 A.M. What great fortune—we have forty-five minutes before we go to formation. We got a new corporal this morning, and we lined up in the dark, thirty-five degrees outside. Our heads get cold because of new haircuts. The thin hoods on our army jackets are not too warm. Our new corporal will be our processor today. In processing we are set up with a new commander every day. This corporal is a female, and she doesn't want us to talk about women. (Tell me, what else do forty-eight males talk about when they get together?) Corporal doesn't want us to swear in her presence. She tells us she loves the army, tells us the civilized world is disorganized. Also tells us she was married for six years (who cares about her personal life?). She also tells us how to answer her commands.

Corporal: "Platoon!"
Platoon: "Ready, corporal!"
Corporal: "Ten hut."
Platoon: "Motivated, dedicated United States Army recruit, corporal. "
Corporal: "Right face."
Platoon: "One, two."
Corporal: "Left, right."
Platoon: "Alpha."
Corporal: "One, two."

Platoon: "Three, four."
Corporal: "Three, four."
Platoon: "One, two, three, four."

Corporal says she never swears, but if she swears she is very angry. She says she is only one around here that ever smiles, and if she isn't then she is angry. Corporal says her mother was military and so was her father. She says she still takes orders from Mom. Says Mom will hit her upside the haid if she swears. Also, corporal talks with a southern accent, as do most of these sergeants and commanders. It must evoke some extra power in their command, because one sergeant who didn't have a southern accent was and seemed like a nice guy.

We went to breakfast and then back to the barrack. At the barrack everyone is sitting on the floor and just shooting the shit. Nobody wants to sit on the beds because if they do, the beds have to be remade. There is no other furniture in the barrack. We are just talking about Basic starting on Tuesday and talking about reveille. This is the earliest that some have ever gotten up in the morning. It's not bad, though, good fresh air.

No TV, great. I am already not worrying about El Salvador, Middle East, or nukes. Ignorance is bliss, I can attest to that.

It's getting light outside. It will be time to leave soon. Today we go to PX and also get our very own uniforms. Everything is done as a group. I really feel like a part of something here, even though I spend a lot of time taking notes.

Just daydreaming a little bit and I thought I would write about respect in the army as I see so far. I have a feeling that rank begets respect and people who crave respect crave rank in the army. Not many people get show of respect like they do in the army. It is too bad that respect is so hard to get in civilized world. Too often it is based on money or good looks, neither of which matters in the army. It is too bad that everyone doesn't get respect in civilian life just by the fact that we are human.

It's 8:30 P.M. Just some reveries and stream-of-consciousness before I go to bed. I am very tired, my sides, hips, and heels are hurting. Mostly from standing all day. I made some notes at the Exodus meeting today and also turned in my money to building 2145 for safekeeping tonight. Officer who accepted money was from near La Crosse and went to UW-L like I did. He was very happy to find out I was from La Crosse and gave me a genuine smile. Some sexual flirtations around among the noncommissioned officers. Mainly Corporal Petit and Sergeant Pryor, a big, cocky sergeant, gruff and 'strutty. Seems like another consummate ass. There was some humor while waiting to turn in money. One black corporal could not speak very well and was acting like a cock. Trying to get some recruit to do seventy-five or ninety push-ups. He also argued about football teams and also ordered one individual around. First one way then the other. Thinks he is neat. Is really another consummate ass.

4 December 1981, Friday

Confound it! No room in these lockers. Stuff is tumbling out. I was told to empty duffel bag with all its contents and bring it out to be labeled. Rest of clothing is a tangled mess and strewn in any way. My laundry bag is tied up to the clothing bar; rest of my civvies hang there. I need at least ten more hangers. I have eight. Got up at 4:30 A.M. today. Someone screwed up. We should get up at 4:00. We don't have enough time. Last night after lights out, 9:30 P.M., the sergeant came in and inspected. He was insulted by the latrine and dirty stairwell. He made us get out of bed and mop floors and get black heel marks off. I had to sacrifice a clean towel of mine for cleaning.

Cold today, northwest winds. Had hood on and earflaps down. Wasn't supposed to. Lined up in platoon, marched to mess hall. Ate eggs, hash browns, orange juice, and coffee. Eating less, more time, tastes better. A-A-A-A-A-A-AH, what ecstasy. Then walk to barrack after breakfast. My screw-off time. Dark, no one can see me screwing off. I walk with hands in pockets and relax. Stomach warm from breakfast. Fresh cool winter winds, thrashing trees. Barracks aglow with warm lights. Today barracks were open before I got there. Usually they're locked and I have to run to get key.

It's 9:50 A.M. We have until 10:15 to get into formation. We are up in the barrack; everyone is dressed in their combat fatigues and sitting on the shiny, clean floor. We are scheduled to get dental X-rays today. Jiminy Christmas, why? This is about the fourth set of X-rays that I've had for the army. We don't need that many. We probably glow by now.

This morning was a two-hour meeting in the reception station hall. Eight TVs set up, with several movies shown. The room was bedecked in the flags of every state in the union. The films were on prohibitions of the army, protocol, military judicial code, and introduction to basic training. Best movie was that of military judicial code. It was about mutiny, selling government property, desertion, not being on boat for shipping, surrender, resisting arrest, respect, abuse of prisoners, and activities unbecoming of an officer, such as, having sexual relations, spying, unlawfully holding prisoners past detention time, murder, and robbery. The film was humorous in its setting but contained very important messages about our rights, protection, and duties.

6 December 1981, Sunday

Beautiful morning, clear, nice sunrise and about thirty-three degrees. Last night I had a great dream. A sergeant kept talking and not saying anything. He kept talking faster and faster and his mouth started flopping and flopping and finally his head rolled off his shoulders and hit the floor. It broke in half like a cheap plastic bowl and was hollow.

Last night I had fire guard duty from 12:30 to 1:30. When they woke me up I had only been asleep a few minutes, drat the luck. My duty was uneventful. I imagined I would save the whole platoon and get a medal. I imagined a fire broke out and I was a hero, running to and fro saving injured soldiers and such. Just a reverie. Guard duty is fun except for waking up and not getting any sleep. The barrack is relatively quiet, with just your footsteps and the creaky floor. You can hear the breathing and snoring of the men and the blowing of the furnace when it is on. Once or twice a soldier will talk in his sleep about his girl or something. The street-lights shine in in eerie ways, and a black and white cat is on the prowl. The knifelike shadows are still and peaceful. How many men have passed through these barracks? Where have they gone? World War II? Korean War? Vietnam? Maybe heroes left here.

Writing from snack bar at reception station. We are on break time. A very well deserved, hard-worked-for free time. Two hours. Being good all week, doing what we are told, keeping barrack clean, dressing properly, and, most important, passing today's inspection. We cleaned everything at least three times. Floor swept and mopped, windowsills, latrine, shiny copper pipes, shower, sinks, our beds, all straight, tight, and orderly. One doesn't really appreciate a free moment to himself until one doesn't have it. We do not do anything without an order. We have to be in bed at a certain time. Usually 8:30 or 9:00. And then a sergeant will knock on the door, roust us out of bed without a qualm, and then inspect the barrack. If anything is out of place or not cleaned properly we must clean it or fix it. We must all be in our beds, we must be covered up; our belongings must be put away and locked up. I am not saying I dislike it or that I think it is a farce. I just begin to get tired. I begin to want a rest from it. I begin to feel like I can't keep it up. But it is ironic that the best way to feel good about it is to be very positive about everything. Jump into every job or task happily, with relish. That is the only way to keep your head on straight. Avoid trouble, enjoy yourself, and earn free time. Free time. I never thought I could enjoy time to myself that much. Never realized how much my freedom means to me. The tight regimentation of the army is necessary to make the army effective. But it is nice to know that civilized life is not like this and that we have freedom. This regimentation also makes one a better soldier in a way of knowing what tyranny is like and makes us more energetic in fighting it. If all of life was like this it would not be fun. Never a choice or chance to be one's own kingpin. Never a chance to have one's own castle. Never a chance to enjoy one's own time. Long may our flag wave.

Morale is high here at Fort Leonard Wood, at least in this processing stage. It will be interesting to see how much morale changes when the real Basic Training starts. Morale also seems high in the ranks of the corporals, sergeants, and other cadre. They laugh a lot, they often carry with them big smiles, and they are enthusiastic about their work. They are among the happiest people I have ever met or worked with. They feel like a part of a group, as everyone here does. I believe they feel like they have a real purpose here and in life, as they should, because they do. Oh sure, I have met one or two sarcastic officers, usually in medical or some other support area. One needs a grain of salt to exist in the army, as in life. But some people seem to have a grain of lye instead of salt, and their caustic attitudes dissolve any semblance of a bond between people. They indulge in sarcasms and think they are being smart trying to belittle some innocent soldier below their rank. They abuse their rank and also hurt themselves. I know that those venomous sarcasms are doing more harm to the giver than the receiver. Fortunately, the vast majority of the people here are respectful and considerate of the feelings of others. The sarcastic people are the ones who sit in their niche and rot. Give respect here and you get it.

I went to Catholic mass today at 11:00. It was a nice service. The priest had an accent that seemed of Slavic origin and was not native to this country. He spoke softly and was not a perfect enunciator of English. The church was small, about sixty feet long, forty feet wide, and twenty feet high. The carpet was plush and maroon, as were the priest's vestments. The ceiling was supported by a framework of black joists and beams, like a bridge. There were gold lamps between every window, and there was also an organ player. The service was Catholic and admittedly it's been a long time since I have been to mass. The most striking item to me was how mostly everyone, girls and guys, was wearing their combat fatigues and army green field jackets and, of course, the shiny black leather boots. About one-third of those attending were wearing civilian clothes. Off-duty military personnel no doubt. I have become so accustomed to wearing and seeing combat fatigues that seeing those civilian clothes shocked me. They were different; they were civilians. They were exercising their freedom, and this brought home to me my difference, brought to me what I am, what I will be, and for how long I will be that. It is rather a melancholy feeling. The service was good and was concluded. The priest slid in a discreet comment alluding to his happiness in being in this country, and that made me think also. As the blessing was concluded, the priest said to us, "Do you all love the army so much you would join in winter and not wait until spring?" Funny.

Writing from snack bar. Many electronic games: Space Invaders, Galaxians, pinball, pool, TV football and submarine.

This sounds like a circus. Whistles, bells, music, explosions, shouting, and talking. Everyone dressed in camo. Looks like the opening day of the archery deer season.

7 December 1981, Monday
    Notes lost.

8 December 1981, Tuesday
    This is a continuation of the many notes I took today. All of those notes are lost, since they were in a pocket notebook and I have no idea where it is. The time is now 5:40 P.M. We are once again on free time . . . God, what a relief. Free time again. Free time and playtime until 8:30 P.M. Then we must be back in barrack and clean until 9:00, then lights out. Tomorrow up at 4:30 A.M. and we ship to Basic, to the Third Brigade. Tonight at 2:30 I have fire guard duty once again, for an hour. I am in barrack 2142 tonight. Some of the people that were held back today from shipping from the other platoons are here, too. All INSCOM people, that is, intelligence community.
    After all of my waiting around today at the INSCOM building, about nine hours' worth, I headed back to my original barrack to pack my bag. The long wait at the INSCOM was not a total loss, because I talked to two or three other people for a few hours and they were very nice girls. At this stage, one has to talk to women when one can, because there is no fraternizing allowed here. But back to my pack. It was very difficult packing. All of my issued clothing, combat and dress, boots and shoes, along with my civvies, in that one bag. It weighs about one hundred pounds now and is hard as a rock. When I try to swing it on my back it more effectively swings me. Now I know where the phrase "to bend over backward" comes from.
    I finally dragged myself over to building 2145, which is

the main office of this reception station here. We were then issued bunks in this new barrack to spend our holdover in. When someone says there is a lot of standing around in the army, believe him. I believe this wait training at this reception station is planned so that we learn patience. I am sure that the waiting that we do here is nothing compared to the amount of waiting there is on a front in a war situation. I really don't believe that there will be this much waiting in Basic or while I am in California for my language training. More likely it will be the reverse: so busy that I won't have time to even write or think. Time will tell.

I believe that on this reception station post there are four platoons, with about two hundred men per platoon, and they process and package about eight hundred new recruits a week here. At least this week they did.

We usually ate at the Third Platoon mess hall, which served good food, but tonight we ate at the First Platoon mess hall. It was a bad choice. I was unsuspecting in the long food line. The sun was setting and illuminated the rippling high altostratus clouds to an intensely bright carmine. And closer to the ground a soft, cool breeze. Quite the soothing condi-tions for fine dining. I was further excited as I drew near the posted menu: spareribs, seafood platter, mashed potatoes, pork chops, beef stew, french fries, fish, cake, and pie. By now I was in a dogged drool and ready. But disillusionment was in the air. All that was left was refried ribs, a gallon of shrimp, a few burnt fish sticks, soggy peas, frozen french fries, stale cake, no pie, no cottage cheese, no lettuce, no nothing. I fought my way through a pork chop that could have come off my boot and some peas. The peas I recognized by their white color. I washed it all down with some Coke because all the milk was gone and thought about dying. A sergeant came in and didn't seem to think that the mess hall was clearing fast enough, so he yelled, "Last four men left have KP!" This was one of the best motivation tricks I have ever seen. If you have ever served on a KP team, you know you try to avoid it. This I did . . . but barely. I was the first one out, but just as I was going out the door the sergeant said, "You five have KP." What a liar, what a skunk! I let the door slam and slowly took evasive action . . . with ten footsteps. I was lurking on the other side of the mess hall in seconds flat and made good my escape. I did not desire one bit to spend my first hoard of free time scrubbing grills, emptying cans, washing floors, and the like. No, they had no right. I succeeded; I escaped. I was free for now. I found out later that the sergeant pursued me out the door but gave up after I vanished without a trace.

Back at the barrack. I am in my room now. I have one of the two rooms at the end of the floor that all these barracks have. These are double rooms with two bunks and two lockers. Not the greatest, but the added privacy is much better then the main commune hall. This room is a little beat up. About eight by ten feet. It has two windows about four by three and shaded by twisted gray venetian blinds. The walls have about ten holes, and there is some graffiti. Some spots are plastered over and unpainted. The walls are the same bluish-green, and boards make up the south and east walls along with 1/4 inch cracks. Those walls with the holes are made of Sheetrock. The room could use some work. However, the beds are neatly made and will suffice. I know these barracks are not the greatest, but compared to the jungles of Vietnam or the deserts of the Sahara or the trenches of Europe these barracks and food are fit for a king. We are not here to learn German or to earn money or to see the world. We are here to stand, defend, and, if necessary, **die** for the USA. Anyway, tomorrow I'll be starting basic.

It's 8:00 P.M.: About one half-hour of free time left. I just got back to my room. I went over to the snack bar, and there were about fifty guys and one girl. The place was very smoky, so I left. I then went over to the PX to see if there was anything I could buy. I was not allowed to go in because a sergeant was taking a roster through to buy personal items. So I decided to walk around the reception area. I have been here over a week now. The area is not too big and I have no motivation to wander off because there is a $150 AWOL fee. I walked up the road toward the First Platoon mess hall, and I met up with Mike and Al. They called to me and so I just went over to talk to them. We just talked about the army and why we are here. We all agreed that we have a very good deal. We do not have to spend money for room or board, and we are able to save at least $500 a month. That's great! While we were talking we noticed a couple kissing. The guy came over to talk to us, and we told him that fraternizing was taking chances with the rules, the way they are here. There is a $150 fine for that, but he said it was worth it. He then said he was getting married. What? Getting married. I then asked him when they met and he said, "Last night." I then asked him when he was getting married and he said, "Next Sunday." The guy was twenty and from York, Pennsylvania. The girl was nineteen and from Orlando, Florida. I suggested that he just maintain an engagement and if the love was true it would last. I couldn't let him walk away without giving him my opinion. He still maintained that he was going to do it. We congratulated him and he walked away, shivering as he went.

Back to some notes about the barracks. Most of these barracks were built in 1937, but this one seems older. The barracks are sided with a light yellow aluminum siding. But depending on the time of day the barracks appear white, yellow, or green. The latrines in these barracks smell like urine in outhouses. They don't give us any cleaning compounds to clean with. All they give us is a mop, towels, and water. This should be changed. In every barrack is a garbage can turned upside down, empty, and with the top on the bottom. We cannot use it. Why it is there nobody knows. If we have any trash we have to walk outside fifty yards to the dumpster.

9 December 1981, Wednesday

It's 7:00 A.M. The drill sergeant, Sergeant Skinner, has been showing his face around now and then. Nobody knows his first name; everyone is afraid to ask, for obvious reasons. He is a peculiar character. He is tall, about six feet two inches or so, lean, and well built. His face is slightly weathered and tan. He has intense dark eyes. When he talks to you he will cock his head constantly at different angles but will always stare you in the eye. His voice is sharp, clear, and fast. His movements and walk are decisive, direct, and fast. He is always wearing a Smokey Bear hat and combat fatigues. They say he spent time in 'Nam. I'm glad he is on our side.

It's 6:30 P.M. We made it to our Basic Training facility . . . finally. We are Charlie Company, Fourth Battalion, Third Brigade. It's supposed to be the biggest and best company on post. As we'll find out, says Sergeant Gibson, the senior drill sergeant. Our other drill sergeants of our platoon are Sergeant Reed and Sergeant Hensen. I will admit at this point I have a lot to learn. From the beginning of this afternoon it has been busy. We waited at 1:00 P.M. at the reception station for the "cattle truck" and were taken to the outfitting building for our battle gear. We were given a bag to carry. In it were a helmet, two pairs of field pants, pistol belt, rucksack, chemical gloves and boots, overshoes for our leather boots, a camouflage cover for our helmet, canteen, a light field coat, a liner for our field jacket, a poncho, and winter mittens with wool liners. All of these were delivered to us in the usual assembly line style following an anti–screw-up lecture. The people handing out the gear were mostly good-natured and not nearly as hostile as the group that handed out our dress greens. There was a man who was telling stories about the gear.

The gear was older than he said. He was a jolly man with a leather face and potbelly. He was telling about the high cost and quality and the fact that some of it was from World War II. He claimed it cost a total of $1200 per man. I didn't believe that, but it is high-quality.

Imagine the stories these clothes could tell. What battles have they fought? What lands have they been to? How many men have worn them? If I didn't feel like a soldier before, I sure feel like one now with these war-worn antiques. I have many strong feelings at this time about being in the army. I am not having second thoughts or anything like that, but I'll admit I am nervous.

After we received our field clothes we were once again herded into the cattle trucks. I don't know if these trucks got their name from the way they look or the way we were packed in. Body to body with our two huge packs of clothes shoved into this semitrailer. An army green big wheeler pulls this trailer. The trailer is silver with some windows. There are benches inside along the walls that resemble park benches. They are, however, to put your baggage over and under. We stand. On the walls is the graffiti of other "cattle" that passed this way, usually reminding us of some soldiers' short time. Many areas are spray-painted black covering up the graffiti, so what we see is fresh art. It's stuffy and dizzying in here. We don't travel far, maybe two or three miles, but we turn many corners and we can't see out.

The drill instructor rode with us, Sergeant Reed, and informed us that this is his last cycle as a drill instructor, which is unfortunate for us, because he wants to go out with a bang. He also informed us that this is the beginning of Phase 1 of our Basic and that we will not be allowed to talk to females for two weeks. That's too bad, because one has been friendly with me lately.

We finally arrived at our barrack and immediately dis-cover where all of the distant shouting we heard from the reception station was coming from. Soldiers are doing PT (physical training)—calisthenics, running, and always shouting. We have to run about three hundred yards with our heavy packs and get into formation. The shouting starts immediately, and if that isn't enough, the DI uses a blowhorn. He called off all of our names, and when we heard our name we would say, "Yes, Sergeant Gibson," standing at attention. One guy messed up and said, "I'm sorry," and Gibson made him do ten push-ups for saying, "I'm sorry." Another girl messed up and didn't say, "I'm sorry," and Gibson made her do ten push-ups, too.

We were moved up to the third floor of our barrack. These barracks rival any college dorm from the exterior. About one hundred yards long, beautiful red brick, and three floors high. Inside they resemble a dorm, too, but are lacking a few things, such as tables, chairs, and bulletin boards. The rooms are big, twenty feet by twenty-five feet at least, with about a nine-foot ceiling. There are roomy wall lockers about six feet high, three feet wide, and about two feet deep. They also have three drawers. They are easily superior to the wall lockers that were at the reception station. There are three pairs of bunk beds and half a bunk, for a total of seven beds. Mike Snyder and Dan Vieths are two of my roommates.

Already the sergeants have been yelling a lot, and already I have made a few mistakes. I took one blanket when I was supposed to take two. I also failed to stop and face Sergeant Hensen. After a meeting we had about our upcoming leave, we had to turn in our money. I got in line with $50 to turn in. Luckily I was first in line.

This place is really something and impresses me already. I can see recruits outside marching around in full combat dress with M-16s, helmets, the works, and they are always shouting. Right now they are yelling, "The outlaw gang, the outlaw gang!" and other things. They look enthusiastic; they look like soldiers; they look confident. I wonder what we look like to them? I feel lower than them, and of course, our platoon looks it. We do not march that well yet, and we haven't gotten used to the DIs yet. We make mistakes, like looking around in formation. We do not know the area yet. We look like, and of course are, novices. Time, work, and our DIs will, I hope, change that.

In the mess hall we must stand at attention in line and not look around. We must shuffle our feet sideways when getting food and keep both heels together when not shuffling. Both hands must be on the tray while walking. We must not be eating or drinking while standing. We cannot drink any soda pop for the first two weeks. And while in the cafeteria all of the sergeants look for errors and shout awfully loud at them. I made a mistake of not having one button buttoned on my coat and had to go sit down. The drills shout and shout, and I think too sarcastically and loud. The drills laugh among themselves. It's true, a sad, hard, real fact of life. These people here are professionals, damn good ones. But it will just have to take time to get used to it.

On the barrack wall downstairs is a terse statement: "An undisciplined soldier commits murder. An undisciplined soldier is a dead soldier."

10 December 1981, Thursday

Got up at 4:30 A.M.; female drill sergeant came in my room last night and threatened to confiscate my alarm clock because it wasn't secured in my locker. Early in the morning we filled out forms for laundry, home address, and leave coming up. Were also shown positions at various commands; they were 1) stand at ease, 2) at ease, 3) at rest, 4) attention, 5) parade rest. Stand at ease and parade rest are almost the same. It's when your feet are about fifteen inches apart and your hands are folded behind your back just above your butt. "Stand at ease" means your face is on the speaker. At parade rest your eyes are straight ahead. "At ease" is a command often given by superiors after attention is called. The soldier goes to parade rest and can scratch if need be, but no digging in your ass. "Attention" is a command given often at the appearance of an officer. The soldier stands straight up with heels together and toes about an inch and a half apart. There is not any movement, and hands are hanging on the side of the hips. On the command "at rest" you can fix uniform if need be, but no moving right foot.

11 December 1981, Friday

Weather this morning cool, thirty-five to forty degrees, great gray, cloudy day. Got our weapons today. Such a proper day to get our weapons. We lined up in our company, about 200 of us. We all counted off until all the numbers were gone, up to 180 or so. We lined up by number to get weapons from arsenal. We must always stand at attention. The drill sergeants look us over and correct us, always yelling. But we get our weapons. My number is 167. My serial number of my weapon is 4969236. I will have this weapon my entire time here. Weapons are M-16s, A-1s and are very light. I expected a twenty-pound lug, but this one feels like a toy. Still, it's no toy and possesses a terrible seriousness. When we get within three men of the weapon handout window we yell out our number and then give up our weapons card. We must always have either our weapons card or our weapon in possession at all times, or we are in trouble.

Sergeant Hensen went over the assembly and disassembly of the M-16 with us for most of the morning. I can assemble and disassemble my weapon in a very short time. In a matter of seconds.

It's 5:00 P.M. Man, oh, man, this is a riot! I mean it, really.

It has a flavor of a comic unfounded. We had one hour of PT today. We were just exposed to the PT area where Charlie Company will work out. A large field. Charlie Company has 240 members, including 40 or 50 girls. Sergeant Gibson will lead with a blowhorn. If a soldier doesn't follow orders exactly, he is ordered to the front leaning rest position to do ten push-ups per offense. Usually an order to do so sounds something like this: "Get down and knock me out ten." "Who, me?" "Yeah, you, stupid"; then down the soldier goes.

When we are all spread apart by five-foot intervals the other drills walk around and catch other people who screw up. It's pretty comical. Most of the mistakes are from not listening, and the ultimate goal is to make us listen better. The only way they know you are listening is to make you do everything they tell you to do. And they give us many small details. I know for a fact that I am listening better at this time than I did two days ago and much better than a week ago. I am still glad I came at this point. It will remain to be seen if I stay that way. I think I will, however. I am looking forward to being a better person for all of this. I missed a bit of training today. About an hour. I and twenty other recruits from our platoon. We had to go to the commissary and buy running shoes for our PT. We will have one hour of organized PT each day, weather permitting. Today at PT we all had to growl when at ease was called. It sounded rather funny and also hurt our throats. I am hoarse from all of the yelling and singing we did today. When we're at PT and Sergeant Gibson yells, "Attention!" we have to yell back, "More PT, Drill Sergeant! More PT! Make it hurt, Drill Sergeant! Make it hurt!" From time to time we have to yell out our company motto:

> "The difficult we do immediately.
> The impossible takes a little longer.
> Miracles by appointment only.

Duty, honor, country.

Drive on, drill sergeant, drive on."

12 December 1981

It's 5:20 A.M. Shit! Last night I had to do more push-ups for Sergeant Hensen two different times. That makes forty extra push-ups so far. The first set was when we were turning in our weapons and supposed to be standing at attention. I had moved my arm a bit, and Hensen said, "Anyone that moved at attention, drop." I was on the other side of the column but didn't want to take the chance that he hadn't seen me, so I knocked out ten. The second set I did last night was when we were over at the PX to buy things and after we were out and at rest. A soldier asked if he could smoke. Hensen said, "You should know better than to ask that . . . drop." Then Hensen asked us, "Who can tell me if it is all right to smoke?" I then blurted out, "Yes, it's all right." Hensen then said, " 'Sergeant.' Now you drop." So down I went and knocked out ten and then got up. He then made me go down again properly. Front squat position to the front leaning rest and then knock out three more, each time counting off and saying, "Sergeant Hensen." I then had to recover like this, saying, "PFC Stanley requests permission to recover to attention."

It's 6:00 A.M. About five minutes ago, Sergeant Wert called out my name. I said, "Yes, Sergeant," and ran down the hall to see why he called me. He then gave me two pairs of GI glasses. The frames are black and heavy and don't seem to have the quality of the usual GI goods. They will take some getting used to. I then forgot to say, "Thank you, Sergeant," and he made me knock out ten.

Today we have a PT test. Push-ups, mile run and sit-ups. We have a superjock contest where the top 15 percent get an

award from the company commander and First Sergeant
Williams. I feel out of shape, but I will do my best.

On the way back to the barrack today we sang a song.
Senior Drill Sergeant Gibson called out the cadence on the
blow horn and we all sang along. It went like this:

> "Tiny bubbles
> In my beer
> Make me wonder
> why I'm here.
> I don't know
> What I've been told,
> That Charlie Company
> is good as gold.
> I don't know,
> But I believe
> That I'll be home
> By Christmas Eve."

It's 9:00 A.M. Just went over procedure for salute to an
officer and how to report to senior officer or sergeant. Sergeant Hensen is very good at giving examples. He is very
disciplined, patient, and impressive. I hope I can do it as well
as he.

We had a meeting this morning with Captain Liter and
battalion commander. It was a boring meeting and didn't
really accomplish anything, but I screwed up three times.
Once for not having a pocket buttoned, once for not saying,
"Sergeant," and once for turning left instead of right. I was
given five push-ups for each offense.

Last night I had a rather humorous experience. I came
into my room after a classroom meeting and found I could
not open my lock to my locker. I tried and tried and still found
it would not open. I walked into the classroom last and
confidently told the sergeant "My lock won't open." He said,

"Impossible." We walked back to my locker and tried again. Still no luck. Then a guy in our platoon walked in and asked us what we were doing. I said, "My lock won't open." He then said, "It won't open because it's my lock." It turned out that I was in the wrong room. Shit!

This afternoon was our first PT test. Everything was much more difficult than I thought it would be. In other words, I am not even in as good shape as I *thought* I was. I was not too disappointed with my push-ups. I did all I could. But of course I wasn't overly proud. I think the highest anyone did was fifty-five. I did forty-three of them. Many did in the forties and there were quite a few in the thirties and some in the teens. One guy couldn't even do one, and the DIs got very angry and called him a Big Mac attack. They said he couldn't eat any more dessert. We all did what we could in two minutes.

My sit-ups were also a disaster. I was very disappointed but did all I could. I did thirty-three of them. I don't know why my stomach is so weak. Both the sit-up and push-up tests were conducted outside on carpets under a large open-walled shelter about half the size of a football field.

The last test was the mile run, and that was the most difficult. I thought I would have breezed through in six minutes flat, but no. My time was 6:44. The temperature was forty-two degrees, with a 10-mph wind and high cloudy skies. It was like a Wisconsin November day. The run was very painful for me; my chest and my jaw hurt, immensely. Don't ask me why my jaw hurt but it did. The best time was about 6 minutes and many did a lot worse—8 or 9 or even 10 minutes. I can see we are really going to get worked now. We are going to be put in workout groups, A, B or C, with A being the best and C the worst. The times were bad, but we had to run with our combat fatigues on. We got to wear tennis shoes, however. They said next time we will run in our boots, but I think they were just trying to scare us. We are all enrolled in the fifty-mile club, too. When we reach that goal we will get a patch.

13 December 1981

It's 5:30, the thirteenth today. Also the first day of our squads, our marching squads and our clean-up squads. There are four squads in our platoon, with four squad leaders. We have thirteen men in our squad, with Calloway the leader. Thirteen men assigned on the thirteenth. That sounds like an omen, but of what? We are assigned the duty of the latrine, and it takes us about ten or fifteen minutes to clean it. We have to mop the floors, wipe the walls, clean toilet bowls, wash and dry the mirrors, wash and dry sinks, and wipe off washers and dryers. I clean around the toilet bowls and am done in about five minutes. If one attacks the work with a lot of energy the work gets done quickly.

Now . . . interruption. We just had a meeting of our squad. I was assigned a new job, washing the mirrors. It seems many forgot what their jobs were and we had to reassign. It will happen again though; people will forget. Calloway did not write down who has what job.

It's 9:30. I am on a five-minute break right now. We were in the classroom sitting on the floor. Man, that makes your butt sore. We went through the loading, cleaning, and immediate action of the M-16. Sergeant Hensen is very proficient with the weapon. He also informed us about incentive management. How we do and work will affect our assignments. Says we are doing real well. Says he is happy with us so far and also says our punishments and rewards will be tailored for individuals. If you like push-ups, then you will receive some other punishment. Or if you like to stay in barrack, then he will make you go outside, or vice versa. Hensen is real cool and good to us and treats us with respect when we do the same.

He has a serious, yet facetious sense of humor, and often his answers to our questions are laced with a wry smile.

Hensen says we will all be different people when we leave here, that we are different already. He says he is always interested in seeing the civilian recruits come in and then seeing these same recruits when they leave, how they have matured, how much more self-confident they are. I know that I have changed already and am getting more confident.

Hensen pulled a dirty trick today. While we were in formation he was calling out, "Right face, left face," and then he called, "Rice face." Anybody that turned right had to do five push-ups.

During total control, the first two weeks here, phase one, we are confined to the barrack. While not working, which isn't too often, a couple of times a day we have a ten-minute smoke break, when we go out in back of the barrack and stand at rest in formation. Those who smoke, can smoke and those that don't, don't. It is a time where we can talk to the DI. We had Hensen today. He talked about where he was from. Winfield, Kansas, forty miles south of Wichita. Says that is where Mary Ann from "Gilligan's Island" was from. One of the recruits said he had some relatives there, and Sergeant Hensen said, "That's all you could possibly have there, it's a very small place."

While talking at rest, Mike Snyder and I discussed the no-talking-with-females policy in effect during Phase 1. He and I both agreed that we have not even had the chance to talk. The girls won't even look at us or even glance. I don't know if they are afraid or just don't like our looks. I haven't even been able to look one in the eyes once and pass a wink.

It's 6:30. Right now I am practicing my about-face, approaching an NCO, and approaching an officer. We practice and practice. I am getting better at it all, but slowly. I cannot keep all of my concentration on it. Tonight Hensen stopped me while I was in the dinner line and I screwed the about-face up. Today I volunteered for detail and ended up cleaning the administration office of Fourth Battalion. We had to scrub floors and mop them and then buff them. The buffer wasn't working too well. I notice many *Soldier of Fortune* magazines there.

14 December 1981, Monday

Last night bed at 8:30, got up at 4:50. Sergeant Reid started morning by yelling at everyone: boots not laced by beds, people not saying, "At ease," when he walks in, people in latrine longer than twenty minutes. Slight tinge of snow last night. We were woken up about four this morning by girls marching outside our window. They must have gotten up really early.

Noon: Very busy, probably will not have much time to finish writing about this morning's events. We marched to auditorium before light and saw raising of the colors. Went into auditorium and had a presentation kicking off our Basic Training. Listened to national anthem, recited Pledge of Allegiance. Listened to a letter written by a soldier who would die in battle. His last letter home. Everything was very emotional. A lot of talk about God and country. Two officers proposed a toast to the long life of the USA, and they then broke the wineglasses. Listened to a colonel give us a pep talk on how much potential we recruits have. Also saw movies about VD and contraception. Very tacky.

It's 5:30. Just finished dinner. Very good. Fish sticks, cherry Jell-O, cottage cheese, macaroni and cheese. Choice of dessert: cherry pie, apple pie, carrot or chocolate cake. This afternoon was very good for the agenda. All of Charlie Company was taken over to the chapel to get to know the chaplain and schedule of masses. The pastor was very congenial and friendly. He had served awhile in Germany and two tours of

Vietnam. He talked to us for about an hour about depression and anxiety. These are similar to each other. He had people get up and say what might cause these. Here are a few of the suggestions: shock of new environment, fear of failure, fear of DIs, fear of yelling, family problems. He has worked in battle areas, so he knows how to help people. The chapel was very nice, too. He described some of the events that we are going to be doing during Christmas.

Today PT fun! We were broken into squads—A, B, C. A is the best, C the worst. I felt I should be in A instead of B. But that wasn't my decision.

We did our PT out in front of our Charlie building. We split into our group and then spread out—Sergeant Gibson got up on his dais and was the one to call the exercises and cadence. He would use the other DIs as demonstrators of his exercises and enforcers of his push-up commands to people who make mistakes, such as slapping dust off pants and gloves without being at ease, coming to attention before execution, or laughing.

Sergeant Gibson really has a riot of a time up on his dais controlling everyone. He was demonstrating an exercise with one of his DIs, and he almost tore the man's arms off, the way it looked. We laughed and so did the rest of the DIs.

Often Gibson will go into a minitantrum and say to someone, "Get down and gimme ten—get down there NOW!" He can really yell when he wants; he's fiery. But with his colleagues he is friendly and congenial. I think he really likes the power he gets when he is in complete control of everyone who is doing the exercise (about 225 people). He told us all to run into our ranks (over his blowhorn), and it sounded like "To the right." He thought we just lacked pizzazz, so he made us do a few up-downs standing to prone position. After that we had to get in our groups again and then yell as a group each time he pointed to us or the other group. We also had to stop when he quit pointing at us. He would point to one group, then another, faster and faster. And behind his business face I could see a real smug look and also a wry smile. Many of us broke out laughing, but he didn't see us. It seemed others enjoyed watching us too, for after we started working out, ten birds (starlings) perched on some high wires and remained there until we were finished.

Our PT then shifted over to the athletic field, and the B group had to run three-fourths a mile. That wasn't bad, easier than yesterday.

After coming back and ditching our tennis shoes, we hiked over to the flag. This time the flag looked much prettier than other times—that was because of the setting. Everything drab army green or dried brown. Black oaks without leaves and just a few small patches of weeds. And then the red, white, and blue colors.

Things are looking better; tomorrow a group of us who aren't going on Christmas leave get to go to the rifle range. We will be taking our M-16s and trying to qualify as Expert. If we do, we will be able to go on guard duty during Christmas vacation. I think Christmas leave will be great here. There will be a few girls around, too. And most of what I saw has been getting better-looking lately. And I imagine that we are getting better-looking to them, too.

Also, Sergeant Henderson was saying how happy he has been with us. He most likely says that to every platoon, but being human, we are very prone to believe it.

15 December 1981

Henderson warned us about getting uncooperative with a peculiar and, I think, overdramatic and detailed metaphor. He says he will give us a lot of rope, but like a man falling from the gallows with a noose, the slack soon is gone and the neck snaps.

Today we had breakfast and then a short classroom meeting. We then went down to Gibson's office, all of us from Charlie Company who aren't going on leave. We were put into a truck—I think a half-ton—that had an army green canopy and resembled a Conestoga wagon. I have heard of air conditioning, but this was very ridiculous—the four-mile ride to the range was long and cold. And we got there nearly frozen—cold feet and hands. Upon arriving, we went into a long pole building with a cement floor to warm up in and have a short class on the way to shoot and aim the rifle. A Sergeant Springer did this. But before teaching, there was a Sergeant Warrens there who more or less held an informal talk with us. This was very nice because it is something I miss very much at this time. We asked him what is going on in the world today, and he said Poland was under martial law, Reagan had ordered all Americans out of Libya, and the economy is getting better, but slowly; talked of his kids, talked of his twins, says his wife can tell them apart but not he; and said he speaks German. We also talked of sports, mostly football. He was very helpful in enlightening us upon what was going on in the world. He had an assistant named Valley. Warrens made a comment that Sergeant Valley was the most difficult person to work with that he has ever had. I thought he was being facetious, but I was to find out otherwise.

We filed out of the trailerlike building and passed by building 4. There we picked up an ammo box each. Each ammo box contained three-to-five-shot clips and two nine-shot clips (magazines). Our targets consisted of silhouettes of men twenty-five meters away. There were about twenty silhouettes in a 300-yard line—a firing line. Our shooting points were parallel to the target line and number. Each shooter had two or three small sandbags to rest his M-16 on. And shooting was done from the prone position, following loading instructions given by the tower—and getting comfortable in the prone position we commenced firing (sixty of us). We had to shoot our first five-shot magazines and stop. I didn't do too bad—I shot very low but got a good group. The M-16 has much less kick and noise compared to deer rifles I have shot. To give you an idea how little kick it has, I would hold my nose against the rear sight to maintain a consistent position when aiming.

Shortly after shooting these first rounds, we were required to remove our targets and then replace our ammo in the boxes, pick up the extra brass, and put away everything. We heard over the loudspeaker: "The U.S. has declared war on—" And then my heart sank. I thought we were at war. But the announcer said, "Fort Leonard Wood."

The colonel has shut down range 5 (because we were not authorized bureaucratically to be there). It seems the drill sergeants wanted us to qualify with a weapon for Christmas so that we could pull *their* guard duty. But their plan failed and they must pull it themselves.

Beautiful day at the range, though—cool (thirty-five degrees), some snowflakes, big, soft, quiet ones. Range was out in the forest; looks so much like Wisconsin, but much warmer for this time of year.

I have been talking to a girl here—not too attractive, a little plump—but she is cute, friendly, speaks French, has not too many friends in States because she lived in Europe for many years. She is 986, too, is going to study Español.

Reid talked about gas mask, talked about how fast someone can move when death threatens.

16 December 1981 (written morning of 17 December 1981)
Just writing a few notes about yesterday:

Very long, mostly classes.
Lecture on military courtesies by Sergeant Gibson.

Lecture by Sergeant Javenex on equal opportunity and
     prejudice.
PT. I did many, many push-ups. Sergeant Byrd kept punishing
     us for the way we marched and the way we got in forma-
     tion, said, "We DIs ain't bullshitting around."
Lined up for march to classroom after supper.
Gibson made the whole company get down in front leaning
     rest position and pump out ten slow ones (five, four
     count).
Went to classroom, listened to Sergeant Reid give lecture on
     male military uniform. Sergeant Utley gave lecture on
     female uniform.

Today, 17 December 1981, it's going to be a long day.
Many of the platoon are leaving. Today is going to amount to
little more than waiting for them to leave. We got up today at
4:00 P.M. Went to bed at 7:30.

18 December 1981

     Writing in room of A Company. This will be our company
quarters for Christmas. Got here at about 1:00 P.M. Last night
I had fire guard at Charlie Company. Got up at 2:30 P.M. and
have been going ever since. No wonder I am so tired—it is
now 4:45, fourteen hours and fifteen minutes. I should be in
bed. I was talking about fire guard! With the exception of
getting up in the morning, fire guard is not bad. It is peaceful.
I can listen to the water fountain click on and off and listen
to the whir of the fan. The light comes in from streetlights.
The hall shrinks as the length of it grows longer—doors
become closer together. Such peaceful shadows. I also read
the sign on the DIs' door—profound and funny. At 2:50 A.M.,
people began waking and getting ready for their travel home.
Before my fire guard time was over, most were up, and so I
began cleaning up. Sergeant Hensen later talked to me about

leaving my post before I was relieved by him. I didn't realize I had to be relieved by the sergeant.

Kawalski again walked into our room, three-plus times today. The rooms all look alike, so if one is not thinking, one can easily walk into another room.

Today dawned very clear, cold, and windy. I would say the wind chill was about zero degrees. When we went to lunch today, we had to hike about a quarter of a mile. And the wind was a stinging knife against our ears. Now at Christmas, many trainees have been put together from several battalions and companies. We have a particular sergeant who is hiking us today, and he really doesn't know how to call cadence very well. As a result, we cannot march very well, and he blames it on us. If we screw up he puts the whole lot of us in the front leaning rest—and we do twenty pushups.

Fall out—to chow.

Colonel remarked that he has spent three years out of twenty away from his family. But he felt that Christmas does not happen in a place but is something that happens inside a person.

Today once again we went to the chapel. We had the Nazarene chaplain, a lieutenant colonel, and a couple of civilians talk to us about what kind of activities we have for over Christmas.

When we were walking up front, a friendly man in green fatigues was asking people if they know why they are here for Christmas. I imagine that most knew why they were here. Mostly to save money and leave time. Some were here because they have nowhere else to go, like Pvt. Lisa Bronas. One man was supposed to be gone because of a medical discharge, but certain things were not done.

Well, this man asked me if I knew why I was here, and I believe I replied curtly that I didn't want to go home. And he said, "Well, that's a good reason." I was impressed by his reluctance to probe and continued on. I thought he was another pastor, but I should have said, "sir." I didn't find out until he got up in front of us all that he wore the oak leaves of a lieutenant colonel. I felt pretty bad for not showing him the respect he deserved, but I didn't see him alone again to apologize.

19 December 1981

I am writing this in my room in my Christmas barrack. Last night an asshole came into our room and started talking: "She-e-et, mo fucker, she-e-et mo fucker." Someone was reading a *Penthouse* magazine, and he kept saying, "She-e-et, Ah don't need no magazine to get off; she-e-et, I done all tha' she-e-et. Ah'm the best gambla' in da barrack. Ah done won three watches and fie tousand dollars already. Ah eben got a gahy's travelin' check in mah wallet. She-e-et, she-e-et, Ah done all dat she-e-et. Ah'm gittin' out a dis ahmy, man, she-e-et. Dey treatin' me like an animal, man. She-e-et, Ah'm a man. Ah used to lif' weights, man. Dey don' treat me lahke a man—sheet mo fucka, I done all dat she-e-et. Ah'm from Tulsa, Oklahoma. She-e-et, man! Ah'm twenty-five and got fo' kids. She-et, man, she-e-et! Ah ben drinkin' every night here. She-e-et! Der's too many dum fuckas hea. I cain't way to git outta hea. *She-e-eit, she-e-eit!* "

Left for chow before sunup—very cold, about ten degrees or so, with wind chill it is probably about ten below or so. Skies are clear and the approaching dawn is changing the black night to maroon to orange and blue.

It's 8:35 A.M. and we just got back into the barrack a couple of minutes ago. We were pulled out into the cold by an angry sarge. It turns out two of the rooms upstairs on our floor were told to go downstairs and police (clean up) the area. The sergeant (Sergeant Price) decided seven more people were needed—he asked a trainee to go upstairs and get another room; the trainee said *no*. Sarge got pissed off (understandably). The Sarge then called everyone out and started swearing at us: "Oh, you fuck-offs, can't you set up a formation right? I'll talk to you in street language then, if you can't understand orders. Break it off—goddammit, break it off! *You* five on the end, *drop;* you hear me, you shitheads, *drop!*" The men didn't drop immediately, and so the sarge picked up a large jagged chunk of limestone about as big as a softball and stomped over to the end of the formation— everyone dropped then and knocked out forty push-ups. (I feel the sarge was unfair in meting out punishment. We were all told to police around the battalion headquarters.)

During reception station, we watched a film about the military—said no one should use profanity. One of the rights on our rights card is not to be abused physically or verbally— or to be cursed. The design is not to disrespect the individual soldier. You can get just about anyone to do anything, I believe, if you can maintain his self-respect.

Tulsa was then forced to carry the big flag with him around while he cleaned up and is also supposed to carry it around with him now. He is constantly running his mouth about; he talks about these white boys. A black roommate accused him of being a racist.

While we were policing the HQ grounds, Tulsa kept antagonizing a recruit and the recruit just kept giving it back to him. The white recruit kept telling him how he was sick of hearing Tulsa bitch—bitch, bitch, bitch. Tulsa thinks he can take on the world; he equates having to take orders with slavery, meant to take away one's respect and not as a form of cooperation, which it is.

Tulsa said to the recruit who was arguing with him, "Have you ever been smashed by a nigger, white boy? Have you? I don't have to do this she-e-et. Ah'm gittin' out of this army."

I was talking to Kawalski, and I said, "If Tulsa jumps on that white guy, we'll be all over him."

Kawalski said, "No, you won't; you aren't that stupid. It's not worth it! Let them take care of their own problems. You'd just get busted, too."

I thought a second and agreed with him and just went back to searching for butts.

Snyder is getting sick again. I don't think he has felt really well since he has been here. He said he has always been in good health but can't understand why he is not feeling well. Calloway remarked, "You're in the army now." I suggested to Snyder he go on sick call, but he said that when one goes there they treat you bad and make you not want to be there. Says the sick room tries to convince you you are not sick. So that is why Snyder doesn't want to go; says Sp 4c. there is an asshole. Snyder is eighteen and from Ohio.

I will start some biographical sketches here. Any story of life and about life is really not that interesting unless there is biographical data that people can compare themselves to. I've noticed here that most recruits here do not hesitate too much to share things about themselves (like Snyder). Some talk about their personal feelings. Some talk about the things they've done (Calloway), and others brag (like Tulsa). Some talk about the places they love: Snyder, Ohio; Calloway, Georgetown; McCubbin, Nebraska; Stanley, America. Some talk of their women: Snyder, Amy; Lamar, his girl and son, not married but supports his son; Vieths, his Jewish wife (his first left him); Barnes, goofy, good-natured Barnes, his wife started running around after eight months; Kawalski shows us pictures of his girl. People sometimes talk of their education. We have many here who have college degrees. That is because most of us in Charlie are INSCOM (intelligence). And since most of us will be studying languages, most of us have degrees. Barnes, who knows Russian, was an electrical engineer for

Boeing; Calloway, who also knows Russian, has a B.S. in psychology; Vieths, who knows German and has a B.S. in microbiology, was living hand-to-mouth in St. Paul.

Lackadoo is thirty-four years old, an X-ray tech, worked for Union Carbide. His friend is a captain and talked him into joining the army. He said he wasn't doing much, had no one special, was making over $20,000 per year, was bored, needed something different. He told me of a girl he met before he left, a real sweetie, twenty-five years old; she was a virgin. She has a master's degree in accounting and is an accountant with the IRS. She sent him a letter and card, and he let me read it. Such feeling, such tenderness, it was a pleasure to read even though it wasn't mine. How much she had wanted an address to write to; how relieved she was to receive it! She had never had a boyfriend before, and this was the first! What a letter like that can do for the receiver and for the people around him. What a great distance that tidal wave of love travels. And its ripples wash to every nook and corner in this swamp. It stirs the deepest and farthest waters.

I am feeling somewhat anxious all the time. I am not sure why. I guess I am starting to feel a little tied up, which is only natural, I guess. I guess I am forgetting what I should do when I get some free time.

I am beginning to know and like some of the recruits very much. I am finding it comes out of a respect for that person's intelligence and conversational ability and sense of humor. Snyder and Kawalski are becoming particularly good friends. We laugh a lot, talk a lot about religion and politics, and have fun. Kawalski likes to imitate Sergeant Hensen by saying orders like Hensen does. One is "Stop moving; stop moving," or "Dress it right; dress it right," or while in the barrack he will say, "Third Platoon," and people will say, "Yes, Sergeant." Kawalski is from Maryland, close to Washington, D.C. He will be studying Russian. Says he has two years of college but credit for only one.

It's 2:40 P.M.—sun streaming through the window. I am alone in my seven-man room. Just a few short notes here—I am on pass from 1:00 P.M. (1300 to 1600 hours) to 4:00 P.M. This is so much appreciated; it gives time to reflect on what has been happening lately and an ability to think about oneself only! I am in the barrack and only about three or four others up here on second floor, where there are normally forty or fifty. So peaceful, no noise—just faint hum of lights, water or fountains. I just got out of a long hot shower. Plenty of time and hot water—um-m-m-m. No anxious feelings about cutting other people short of water like there usually is. No fifteen- or twenty-person shower line. Oh, so peaceful! I really love my free time so much! People who have never been tyrannized to an extreme cannot possibly understand the extreme sacredness of one's freedom and personal time. So sweet, so sweet. Well, I am off to the dayroom and library now; will write later.

I walked over to the PX. Such a beautiful December day! Clear skies and forty-two degrees. People are Christmas shopping and buying Christmas trees. And Christmas music is playing.

I became aware how militarized I've become. I noticed one young man, probably a dependent of some military personnel. He was wearing an army field jacket with Levis and shoulder-length blonde hair. He looked so undressed and out of place. I would not have thought twice of that outfit before I entered the army. In fact, I often wore a field jacket and jeans myself.

At supper this evening Snyder said that being in the army is making him prejudiced. He said all the blacks he met are always jive-talking—street talk (except Lamar, he does sometimes, but not all the time). One of my roommates today corrected a black (also a roomie) about his opulent use of the phrase *mo fucka*—he uses it about every third sentence. My white roomie just asked him to refrain from using it so much because he was offended by it. The black roommate got all pissed off and said, "I'll talk the way I want to talk—I won't call you a mo fucka, but I'll use words the way I want to!" Several people around here, I have noticed, like to say "mo fucka" a lot—I myself have never met a mo fucka. There must have been mo fuckas galore where these guys came from.

Lackadoo also remarked he has had enough blacks already for a long time.

We have a guy here named Orfino. He is from New Jersey. Enlisted in the army reserve to be a mechanic. He is twenty years old, has an Italian complexion and Italian eyes. Kawalski feels Orfino is a lone wolf here. He injured his ankle while running around the track. He also wants out of the army. He says he can't adjust. I ask him why and he just says he can't adjust! I have found in my life that often what seems inability to do something is really not wanting to do something. One is justified, I guess, in not wanting to do things, out either of extreme boredom or physical incapacity. I talked to Orfino about why he wants to go back to Jersey, and he said to be with his friends and so he can party. He could make new friends here and also party, but he doesn't realize that. There are certain things in life one has just got to experience himself. And all the reasoning in the world won't convince him. I remember when I moved to Sturgis, South Dakota, for my soil scientist job. It was okay for a while. But after six months there, I began to feel like Wisconsin was gone forever and I had to go back now or forever lose it. Three months later, I quit and moved back to Wisconsin. It was still there. But I had to prove to me that it was. Maybe I felt I hadn't used it enough when I was there and I wanted to save a happy spot for it while I still could. Perhaps that is what Orfino is going through now. He may feel he is going to lose his New Jersey and his friends forever. Amateur analysis? Probably.

Yesterday, 18 December 1981, we, the holdovers from Charlie Company, received a "talk" from First Sergeant Williams. He told us what we would be doing over Christmas. Told us on mornings we would work and our afternoons would be free. He asked if there were any carpenters in our bunch, and three (including me) raised their hands. He expressly said, "Outstanding!" It is so nice to know when you impress people whom you really like or respect. He lectured us! Today, he said, he was "Father Williams." He said and wants us to be good while staying at Alpha and not to forget *anything* we learned at Charlie Company. He wants us to talk in the respectful tradition to NCOs and officers. He wants us to take care of our beds and belongings to maintain the pride of Charlie Company. Says if we look out for ourselves he will look out for us in case of unfair dealings. Respect begets respect.

I will say that I have a lot of respect for Williams already. He is neat, articulate, confident, and intelligent. Fair also! Says if we fuck up, cause trouble, screw around, or do other actions unbecoming of soldiers, he will come down on us hard! I can see he means it. From what I have seen, the men here in the holdover barrack are different in a good sort of way from the other companies.

20 December 1981, Monday

It's 8:00 A.M. Today will be the first day of detail. What will we do?

- Got out in formation about one and a half hours ago. Glare outside—must have had freezing rain last night.
- Walking back after mess, getting lighter outside, but not bright, rather dingy, heavy dark gray skies, streetlights shining off the streets and pavement, warm breeze on our faces.

• Kawalski and Calloway and I were leaving our dorm. Kawalski and I each stepped on the slick curb at the same time, and we both toppled to the ground. The drill sergeant laughed at that and so did Calloway and so did Kawalski and I. From the conversation around, many of us are falling down. One girl hit her head and got a goose egg. It is rather humorous to fall if that is all that happens and nobody gets hurt. But the glare is dangerous.

Now in my room. There are about five trainees in my room, looking out the window laughing. A few (ten or fifteen) recruits are outside just running up and down the sidewalk and sliding on their backs with legs and arms going everywhere. Some are even bowling—some act as the balls, and others serve as the pins. For they have little time to play here in Basic, and this has been their first real and natural playtime. The infection has spread! The good nature of this extensive ice rink has captured the childhood memories of the drill sergeants, too! One DI came out of the first sergeant's office singing! Yes, singing a Christmas carol—swaying his left hand as if directing a caroling group and carrying a steaming cup of coffee in the other hand! We are somewhat suspicious of his turnabout behavior (for usually we must be serious around them). Then he asks, "Whasa matta, ain't y'all got the Christmas spirit?" We keep quiet—but he doesn't realize he has made us smile inside.

Then we await our morning duties in the supply room, and what happens? The first sergeant is smiling. What style of trickery is this? They must be waiting for us to screw up our protocol! We remain solemn. The first sergeant then brings out fifteen dozen chocolate doughnuts, two strawberry creme cakes, and one huge carrot cake! To top it off, a big pot of coffee! But we are not to be fooled—we move about with shifty eyes, waiting for the curtain to be drawn, the masks to be removed, the lights to be turned on!

But they never did! Not all morning! We were assigned duties, of course—not anything worse than cleaning our own barrack! Just to count a few blankets, mop, and buff a few rooms of company headquarters. And did these in a well-fed state! Another surprise to this army—a pleasant one at that—11:15 comes and we are released to our barrack to prepare for chow.

21 December 1981

What is this! What kind of tomfoolery can this be! In creation's name, what's happened—our barrack, our rooms, our hallways! Garbage cans upside down and emptied of contents! Beds and bedding strewn all over the rooms, garbage everywhere, Coke cans, tissue paper, underwear, dust balls, Kiwi shoe polish cans. The mess took more miserableness to create than what it generates.

Why? Why? What did we do to deserve this? I'm hurt; I'm disappointed; I'm *angry*—loud deep voice bellows down the cement hallways, "Don't touch a thing! Don't clean up a bit! I want everyone to see it. Git down and git in formation!" It's Sergeant Price, mild-mannered, even lovable, Sergeant Price. He must think this a work of art (for everyone to see). I agree it is a masterpiece. No novice could possibly work to this detail. I can only but wonder how many hard years of practice it takes to reach this level of achievement. What kind of hatred he must harbor to have done this! What entanglement his mind must be in! What anguish and pain he must be in to have done this. I see the mess; I do not care! We can clean it up again. We have time enough in this army. All we have is time. We are at their disposal anyway. What does it matter if we have to work one place or another? We do not get paid more or less for it. An artificial or natural mess makes no difference to me, since it is not my time used to clean it. So what? My pleasure in this comes from the knowledge that this mess could not possibly make anyone as miserable as what amount of miserableness it took to make the mess.

We go outside! Price begins his lecture! He begins by saying, "Who has seen the mess?" A few raise their hands. (He gloats.) He says that is nothing to what he can do—says the barrack is a *sty* (pig sty). Says the garbage is supposed to be emptied every morning, says beds should be made to army specs, says floors and halls should be mopped and buffed every day, says latrines are pigsties and will be scrubbed, says someone destroyed a partition in the bathroom, says someone broke a urinal, says someone broke into a display and stole a marksman badge, says someone broke into a display and stole his personal hand exercisers, says someone went through his desk and ate all his candy, says someone went through his desk and went through his personal papers. I was right. He is pissed. Says the badge and hand exercisers *will be* in place by tomorrow morning, or all hell will break loose.

Says he can make a worse mess, says next time he will take all sheets and beds (not just some) and pile them in the center of each room and sprinkle them with scouring powder. Says he will take grease pencil and write on walls and floors to tell us where cleaning is needed. Somehow the good congeniality of the morning is conveniently destroyed. The weather changes, too; the cool brisk morning is melted away, and in its place comes the classic Missouri weather. I knew Missouri could not carry on this fraud of pleasant weather long. I knew the true nature of Missouri before I came. I would have felt cheated not to have had some unpleasant memories of it! Isn't that how this place got its name—*Misery?* People come here to be miserable so that when they leave, everything else is so fine. It was a truly wonderful miserable day, low, heavy gray skies, a fine chilling mist that would moisten your face and clothes. If one played long in this, he would catch his death of cold. When I run, it soaks my hair and eyelashes slowly; then the moisture would drip in my eyes, blur them; the moisture would also land on my lips, and I could taste it. Oh, how miserable it can be. I ran until I could run no more and then walked a bit and returned by running.

The rest of my afternoon until 4:00 P.M. was spent at the Walker Recreation Center. A beautiful building, half library, with many books of literature, and half recreation center. All carpeted, there are side rooms to listen to records or play musical instruments (all soundproof). They have pool tables, pinball, foos tables, electronic games (Space Invaders and Asteroids). They have a large commons area with about twenty living room arrangements—a stage on one end and a large video TV at the other. A couple people are in them. It is quiet, and even the video/audio is muffled. I sit in front and sink into a soft chair and half watch the soap opera "Texas." It is wonderfully boring; I want to go to sleep! A girl walks in; she is cute, slim, and has a Roman nose. I recognize her as one of the girls of Company C—I have not had a chance to meet her yet but do now. Her name is Patty! She graduated from college in '77, is from Boston, has a science degree, worked for a travel agency, and traveled a bit. She has sky blue eyes and a pleasant, attractive figure. She is talkative. I give her a Tootsie Roll. She talks of herself, I talk of me. She *is* pretty. I hope to talk to her again. The time is 3:30, and we must get back to the barracks by 4:00. (Our free time is 1 to 4.) Patty and others decide to take a taxi, at eighty cents a head. Another guy and I decide to run the mile back. I throw Patty another Tootsie Roll as I leave. We run back in ten minutes. Patty is back before me.

Am writing about yesterday, our free day away from the post. I left at 1:00 P.M. with 125 trainees on three Baptist buses from Lebanon, Missouri. The thirty-five-mile ride took over an hour because one bus had a poor clutch, and we rode most of the way on the shoulder of I-44. We traveled over hills of forests and farms. Not big hills, but big enough to be unpleasant to a bike rider, small enough to not bother a hiker. And just enough for pleasant scenery. I did not realize there was so much forestland here in Missouri, miles and miles of forest. From what I saw, there is mostly oak. I would say it is not too different-looking from many landscapes in southern Wisconsin. The land is mature; there are stream valleys with bluffs, pretty.

Cars passed us, limos, Volkswagens, Grand Prixs. Where were they going? Who was driving them? Such a different world inside the cars versus those in the buses—they were so free; we were not.

We got to Lebanon, entered in church, and received an appealing lecture from a Baptist minister about our Savior. We were released to some of today's activities; Snyder, Kawalski, and I went bowling. Lebanon is a dry town, but in the old bowling alley the coffee is good. Some bowling alleys didn't work, so we switched lanes after the bowling machine knocked down all of our pins.

22 December 1981, Tuesday

A wonderfully dreary day at Fort Leonard Wood, a cold, wet, drippy, cold-forming weather.

My lowest expectations of misery have been fulfilled. Any state that has such a dismal name as Missouri has got to have some great bad things about it.

The rain has been pouring since about 10:00 A.M. this morning. It is about 2:00 P.M. now.

We marched to the mess hall today in the pouring rain, and I felt like a dog. We had hamburgers again, the tenth time since coming here in ten days—they are called rat burgers, but I am not sure why. I have not found any hairs or feet in them yet.

While we're waiting in the rain, Sergeant Barker asks who is the highest-ranking person in the platoon. Someone yells, "You are, Sergeant!" Sarge says, "Besides me, Shrivel Brain."

While going in the lunch line, someone said, "Burgers again." Sarge said, *"You can have it your way,* just like at McDonald's." I said, "At Burger King you have it your way." Sarge says they are the same thing.

Today I am glad about five nerds are leaving us, yes, getting discharged—thank God. They are such children; they can't stand still in formation, and yesterday two of them broke formation and ran around a small building and urinated. Why they are all getting out I'm not sure.

Tulsa has two Article 15s, two AWOLs—this morning two MPs picked one up. He wants out, and speaking for the army, the feeling is mutual. But Leavenworth will be his home for a while. One recruit just can't handle it, so he is going home to Mom. Another shot an M-16 in the direction of another recruit at the rifle range. There are no questions asked by the army in such cases—bye-bye, crackpot!

If anyone thinks that the army takes substandard person-nel, he has another thing coming. The army won't stand for it! It *can't* stand for it.

23 December 1981

It's 5:45 P.M. What a fucked-up night! Once again con-fined to our barrack like a bunch of enemy prisoners. I take back all I said about not having any incompetent personnel in our army. We have indeed. This sergeant is a coxcomb from the word go! I realized that from the start this morning when he began taking roll of our platoon. He was practically il-literate, spelling many names he couldn't pronounce and mispronouncing many more. He took the cake, however, tonight. He closed the dayroom on us, our sole source of recreation at night here at Alpha Barrack during Christmas. As I said, our Christmas is formed from many company hold-overs of Fourth Battalion, Third Brigade. There are five companies in training: Alpha, Bravo, Charlie, Delta, and Echo. We have about 130 people from all of these companies here over Christmas. In the dayroom we have three tables, a has-been stereo with eight-track tape, FM radio, and turntable. This morning our company commander First Sergeant Williams pulled out a stereo box with about twenty eight-track tapes that he had recorded—says he hasn't used them for three years. He says *we* (Charlie Company) can have them to listen to and to keep, says he doesn't want to see them again. So *we*—Charlie Company—have our own tapes. Well, tonight at 5:00 P.M. the sarge (the dumb one) reopened the dayroom. Thomas, a private in our platoon, removed the tapes from the dayroom for safekeeping. Earlier today Thomas had tested them on the player and had forgotten them in there.

Well, somebody (not from Charlie) started asking where the tapes were, and the sarge heard him. Sarge thought the tapes belonged in the dayroom and immediately flew into a rage and started yelling for Thomas. Sarge did not know the tapes were for Charlie. He thought Thomas stole them. At *that* moment Thomas began coming back from our barrack so he could return the tapes. Sarge closed the dayroom—and grabbed the tapes from Thomas, and Sarge headed back to the barrack with the tapes, everyone pleading for him to open the dayroom. *But* his mind being made up—and not wishing to be looking indecisive—he would not listen to reason, and hence we cannot go to the dayroom. The dayroom is not a Satan's den, by any means, but it seems like it is the way the cadre talk about it. It has three wooden card tables, two TV sets up on the walls (black and white sets), blaring fluorescent lights (must be where the name dayroom comes from), a white hard tile floor, light blue cement block walls, and about a fifteen-foot ceiling. The doors open immediately to the north wind, and if there are any heaters in the place, I haven't seen them. And *now* it is off limits! Our luxury, which could serve as a torture chamber, removed—we are despondent! But Sarge is right and that's final. Tonight we also have a church-sponsored roller skating party preceded by a prayer service. This too we cannot attend! Sarge says we can't because the dayroom is closed and everybody would just be going skating 'cause the dayroom is closed and not because we wanted to go skating or to a prayer service. Such is the reasoning of an idiot.

The pastor for Third Brigade had us attend a big meeting before the holidays started to tell us of all the activities we would benefit from. What an absurd fraud—what a joke! We haven't been able to attend one. The army thinks we are kids. They are afraid we will break rules! How can we even break rules if we don't even get the chance? Tomorrow I am going to the pastor and plead, for he has common sense and a commonsensical friend who is a colonel. I hope this colonel can put a stop to these childish atrocities that these childish sergeants and sergeant major are putting us through.

God, do I love liberty and democracy! The folly of a life of tyranny is apparent to those who exist in the military. Rightfully, the army requires a rigid system to exist as a functioning unit. But a world organized like this? The system itself would not be the curse, but the fools and tyrants who rose in it would be. Reason would not rule with these people, and their actions would soon show it. We would be subject to hypocrisy and folly in grotesque proportions. This example of the sergeant who would not change a rash decision for fear of losing face would become the rule of a tyrannical system rather than the exception. If one has no power to change something, then how can he?

Very nice day today, after the storm of yesterday (cold rain, snow). Today was too pleasant for Missouri—I have confidence it cannot last long. Mark my words here, for this is *Missouri*.

Anecdote: Today we have a very congenial guy here in Charlie Company, Rod Lackadoo, thirty-four years old, enlisted, quit a $23,000 a year job to join the army, is enlisted as an X-ray tech but wants to switch MO to 986. He scored 122 on D Test (language test).

Since he is the oldest, First Sergeant Williams talked to him today, saying "Lackadoo, I would like you to be a 'father figure' to some of these recruits! Help them; teach them; you know, be concerned about them." (There are twenty guys here and five girl trainees here of Charlie Company over Christmas.)

Lackadoo said, "First Sergeant, I would rather be a boyfriend figure to some of these recruits."

Williams smiled. "Aren't you a little old for them?"

Lackadoo, also smiling, says, "I ain't dead."

Williams laughs. "I'm serious about this!"

Lackadoo laughs back. "So am I!"

Our first sergeant, Williams, is a very fair person, very concerned about our development, but requires us to be disciplined—he is very professional.

Lackadoo got his locker locked up tonight—he is sitting around and talking to me. He unlocked his locker and stepped into the next room to talk to Kawalski a second. When he walked back in, a man in civilian clothes was putting a lock on his locker.

Lackadoo then said, "What are you doing?"

The man said, "I'm locking your locker."

Lackadoo: "Why?"

Man: "Because it's unlocked."

Lackadoo: "Who the fuck are you?"

Man: "Come down to the CO's office and find out."

They walked to the CO's office. The sergeant's face was full of freckles, and he looked like Howdy Doody.

Man: "Hey, Sergeant, I think we have a smartass here."

Lackadoo: "No, I think you got a mad ass here."

Man: "Do you know who I am?"

Lackadoo: "No, I don't. Who the fuck are you?"

Man: "I'm the first lieutenant. You should know who I am."

Lackadoo: "Do you think I'm psychic? How do you expect me to know you without a uniform?

The man then left and took the keys with him. Lackadoo was then sent to clean up the dayroom.

About 11:00 P.M. the sergeant came up to get Lackadoo, just yelling in every room for him. The lights had been out for nearly two hours, and most of us were asleep. The sergeant walked into rooms, turned on the light, and yelled "Where's Lackadoo," until 5:00.

25 December 1981, Christmas Day!

I am thinking seriously. I think this must be the first Christmas I have ever spent away from home. That seems rather hard to believe, but it's true. Today is very fine, fifty degrees. I am in the dayroom; it is 2:18 P.M. The Three Stooges are on. (They are dressed up like generals.) In the corner of the room two GIs (one girl and one guy) are fraternizing. Besides the strict rules about that, it goes on very much here. When one or two want to fraternize (neck), they will do it and they really do not care all that much where they do it. Any shaded nook will do, even if people are around. The only personnel that seem to mind are the DIs, the sergeant major and the first sergeant. If this is what women consider sexual harassment, then they love it. There are about twenty girls in our Christmas platoon, and most have some guy hitting on them. I'll confess also to the charge. It seems to me that sexual harassment only occurs when one person is being "come on to" by someone he or she doesn't particularly like. Even one of the sergeants is fraternizing with one girl trainee, and neither seems to mind.

Today we had a Christmas dinner at the mess hall. It was very good: ham, turkey, stuffing, gravy, pies, shrimp; a very good meal. Patty P. came down and sat by me: that makes about three times. She was supposed to come to the dayroom later but as of yet hasn't shown up.

Last night we had a Christmas Eve party here in the dayroom. It lasted from 4:30 to 11:00 P.M. We had cake, Pepsi, coffee and cocoa and sandwiches and cake and other food. Many of our battalion were here, but some also went to the brigade chapel party.

We mostly played cards all night (Liar's Poker and Spades). I also danced a bit. A few had some alcohol but were not supposed to. Lackadoo, amazingly, got accused of having alcohol. I guess someone poured him a drink, but he didn't supply it. All I can say is Sarge is going to be pissed.

Yesterday I walked to the post office and, unhappily, found it to be closed. I then walked over to the Walker Recreation Center. That was all a very long and relaxing walk. One thing that can be said about the training areas of Fort Leonard Wood is that there are great large fields. When walking these fields from one port-of-call to another, it gives me a by-the-way nature hike. They are mowed grass fields and remind me of the peaceful wide open spaces of the South Dakota high plains. They have a little topographical relief and are pleasant to look at. Scattered here and there along these half-mile-wide fields are a couple four- or five-acre oak groves. These are usually thinned out to resemble a park, and indeed they do, for interspersed among the trees are picnic tables and grills for any who wish to use them.

I find myself at times becoming very fatigued. I am not sure why I get like this. I will lie down on my bed and just nearly pass out from the stress. Well, it happened several times this week and it happened again tonight. It is these black soldiers that are causing it. This whole thing is unbelievable. They live up to stereotypes wonderfully well. They have the loudest mouths I have ever seen or heard. I do not understand why. But when three or more get into a room usually they start up. No matter what they say, it is at the 100-decibel level. And they will laugh and sing off-key, and talk about vulgar things with a vulgar dialect, lips flapping endlessly. They will always try to outyell each other no matter what the topic. They will talk of bustin' someone hard, busting someone's ass, or busting something else. It seems they are always trying to provoke someone to do something. It seems like they have a chip on their shoulder, and it makes people nervous in general.

It is lucky for me that at times like this I can retreat into the recesses of my memories. I can go back to some forest clearing in northern Michigan in the fall—cool, clear morning with the red and gold and brown leaves of autumn floating to the ground, a wind rustling the dried leaves on the ground. I can think of a dark cedar forest, a still, silent solitude, no mosquitoes, no movement, just the cool morning breeze on my face. I can remember windy November mornings on the northern shores with the whitecaps rushing in. All of this mellows me no end. It is not much, but all I have is a memory to think of.

My other roommates think of me as a scrooge here, because when everyone comes in and starts yelling or screaming I will roll over on my bed and pull the folded blanket and pillow tight about my head. If the noise gets worse, I pack up a book, pillow, and notebook and move out to another part of the second floor. And they think of me as a scrooge then. They do not understand me. They do not realize I just cannot takc all of that noise. There are other places for that, the dayroom. I don't know; maybe they are right.

The other night when we were suspended from going to the dayroom or roller-skating party, we became creative. We had a camera that developed pictures instantly, so we put it to good use. We would get in all sorts of smiling positions and sometimes half-naked and standing on each other. Then we would throw out large bursts of laughs, and then the camera would snap. I really let loose that night. Some said they were surprised that a "subtle sophisticate," as they called me, could come that unglued. They don't know it, but that was nothing. If I really became unglued, they, and probably everyone around, would know it.

Within the last two days we have lost about twelve soldiers to discharges, some for medical discharges, most for Article 15s, AWOLs, and just plain wanting to get out of the army. One stuck an M-16 muzzle in his mouth at the range. Most were young, eighteen to twenty. The resultant peace that has come over the rest of our platoon has been great. Now, with the exception of one or two who still haven't been discharged, those who remain wish to be here and act accordingly. Some of those discharged were simply kids, never had the chance to grow up and didn't want to yet. It is good for the army that they are discharged. The army is willing to give many one chance and even a second or third one. I believe that at this time, with our economy and way of life, this is probably the best option a young person has to grow up, travel, meet new people, and get education and experience. But besides all of these benefits, the individual must put in a bit of his own time and concentration.

26 December 1981

The day after Christmas and another terrific lazy day. We were given free use of the dayroom at 9:30 A.M. and free time from 12:00 to 5:00. I pleasantly spent the first three hours of the afternoon in the dreamworld of my mind under a soporific drunk. Around 3:00 P.M. I got up and polished off one hour's worth of reading in *Innocents Abroad,* by Mark Twain. What a masterpiece of literature. Today I ended where he just left the Holy Land. Any and all who are interested in travel will enjoy this work. I could relate more of it, but there is no need to at this point.

A new CQ (Charge of Quarters) today in our Christmas barrack; the prophecies have come true for some and the warnings for others. Sergeant Hensen of Charlie Company is here. It is a relief to see a familiar face, I say. Not only that, but it's a welcome sight to see those reckless and careless vagabonds of other platoons singled out and punished for their childishness and disheveled ways.

It started some after Hensen's duty arrived at 8:00 A.M. He inspected our barrack to see if we swept clean, mopped completely, and buffed lustrously. Any who did not shout, "At ease!" upon the sergeant's entry were soon doing their morning PT. I am gloating now because now the rest of 4-3 is getting a taste of Charlie Company's superiority.

What fun it is to see others punished! Tonight at our 5:00 P.M. formation Hensen was his usually unhurried, precise, shifty-eyed, staunch professional self. Some of the recruits (one-third) missed the 4:00 formation for chow. Hensen smirked a bit at this bit of piteousness. They must go without chow this night. It does not matter too much, though; most are a bit tipsy and will promptly go to bed.

Hensen is a fear-generation (spooky) person—he never yells; he simply tells. When someone does something wrong, he stares at him a second and just slowly, low and cold, says,

"Drop" (meaning to the front leaning rest, i.e., push-up position). There is no tinge of guilt in him or tinge of humor. Something, however, expressly says, however, that you'd better do the push-ups.

While in formation, a guy was spotted by Hensen for moving a bit and told to stand alongside the formation. Originally sentenced to ten push-ups for moving at attention, he was subsequently charged, tried, and convicted in short order for: being drunk, chewing candy, smiling, having a button unbuttoned, not saying "Sergeant," and, of course, general principles. The usual retribution of ten push-ups per infraction was meted out, and the last I saw the poor offender, his arms were not doing anything too active.

L. Bronas is falling in love, I believe, with a Puerto Rican pro basketball player. It is fun to see, I guess, but I can't feel but at a loss in a certain respect. They can talk to each other ceaselessly and are always giggling and cooing about. I never see one without the other. They play pinball together, run together, eat together, watch movies together, and of course smooch together. Whatever else they do, however, is conjecture.

If I was an easier conversationalist, I suppose I would be in his position now, but I am not. She and I would walk together and nudge each other when close and sometimes talk. He, however, is a much better talker, a better listener, a better companion, more appreciative, more interesting, more secure, and more skilled in hobbies than I. He is bilingual (Spanish and English) and probably physically attractive to a woman. I doubt if I lost a friend, and I will see her for a much longer period of time, the duration of Basic, and of course the long stay at Monterey. Let this be a warning or perhaps an example to some, that if a woman is interested at a particular time, then if you are, too, you'd better show it. I did not play games, however, and I even felt like this was about as deep a friendship as the likes of me could expect. But I cannot help being amazed at the rapidity of the formation of new relationships with another of the opposite sex when there were previous flirtations with another. The shortness of time defies expression. I wonder if the relationship turnover rate is so fast in the real world also, but then we do not see it because we are never in each other's proximity for as long, either with guys or girls. In the outside world, a guy meets a girl and when the relationship ends it may be the last time you see the guy or the girl. They are then removed from sight, and you do not continue to see their lives progress. In these close quarters, whether there is a breakup or not, the same people are always in full view, along with who they are affectionate or not affectionate with. Also, there is the possibility that many more lonely people are concentrated together here, and some are able to meet someone of the opposite sex. The odds are in the girl's favor, however, since there are about three or four guys per girl. It is hard to understand that wherever I go, there are more men than women: work, school, the army, night spots, beaches—everywhere. There are always more men than women, not just in one place or two or three. This is not my imagination either. There are *always* more men! I wonder why this is so? I hear that more men are born compared to women, I think about 104 or 105 men to 100 women. I am not sure if this is true. For where are all these women? Maybe it is not true about the odds. If the statistics are true, then there must be many, many girls staying home or hibernating somewhere.

27 December 1981, Sunday

Last night, King Kong came into our room, sauntering in slouching. He was looking for someone to peel like a banana. I turned the lights off, and he grunted loudly, "Who did dat? I'd gonna breaka yo ais."

I said, "Why the fuck are you doing this? Why can't you

ask nicely who you are looking for? Can't you read our names on the front door?"

King Kong: "What if I cain't?"

Me: "Then I think you should learn."

King Kong: "I'm gonna bust you ass."

"Well," I said, "I'll smash your head someday with a brick."

King Kong: "C'mon, man. Do it now, man. Bust muh face, man."

Today at formation, King Kong could not keep his mouth closed. I don't think he had the brains to close it. He just lets it hang half-open and slouches around. He doesn't smoke a cigarette but puts it in a crevasse of his big lips and lets the smoke ooze up into his volcano-size nostrils. He pricks around every day and makes enemies with his fight-picking talk.

Revenge—I went to play Spades in the dayroom! King Kong had the guts to ask if he could play! I couldn't believe it! Of course I said, "No! Nobody can have the last game, especially not you!" He walked away slightly perturbed and dejected. Sweet revenge!

He came back a bit later and watched us play cards—occasionally he would shuffle dazedly over and try to look at my cards and I would cover them. I was trying to spite him as much as I could and was succeeding.

After a while, he even said he was sorry about last night. I was astounded. I didn't even think he could remember that far back in time. He excused himself, however, in that he said he had gone into everyone's room and turned on the light! He said he had lent something to someone and forgotten which room that person was in and had also forgotten his name. I said, "Sure, but you were still wrong, and you shouldn't have had such a bad attitude about it either." I might have believed him, but it was the second time he had done that. One night he was on fire guard. He was supposed to wake up a certain person—our names are on our rooms—but somehow he thought it necessary to wake up everyone and ask, "Was yo name? Who you is?"

(Most of the preceding paragraph was written about last night on this night, 28 December 1981.)

28 December 1981

Got up for breakfast today at 5:00 A.M. Did usual cleaning—went to breakfast at 6:20 A.M. under forty-five-degree clear, starry skies with morning dawning with a rosy-tinted horizon.

Ate chow for approximately thirty-five minutes. Here I was treated to some of the rapidity of the onset of "Missouri." After the thirty-five minutes that we were in the mess hall, we went outside, and a chilling fog had rolled in, from where I do not know. All I know was that it was there. And the rest of the day continued generally in the same manner but grew colder, and the fog lifted.

I wrote a letter to Linda Norvold today and that will take care of some of today's notes.

Went to free church-sponsored bowling retreat today. Bowled 141, 135, 136. Kawalski 141, 221, 204 (top).

Tonight my roomies informed me that we have a homosexual in our midst—said they caught him staring at them in the shower a few times, said he would walk by and stop and look. They told me who it was, the short thirty-four-year-old guy with crutches who smiles at me a lot and says hi. He did seem a bit strange. He is short, very friendly, and does smile an awful lot. I then went into the latrine to take a shower. As I entered the latrine, I noticed the homosexual leaving and I felt fortunate that he was *leaving.* So I went in to take my shower, half-expecting that he would be popping into the shower. The shower ended and I began drying off! I felt uncomfortable for the first time, and so I just faced the wall! Dried off a bit, wrapped the towel around my torso and then returned to my room. When I was in the room, one of my roomies was back in the room and said he was in the latrine for a minute. He said while he was in there the homosexual had returned and walked by the shower room! He said as the homosexual walked by the shower room, he saw that my back was turned and then stopped and stared at me. Upon hearing this I couldn't help feel a bit uncomfortable. His staring is definitely abnormal and understandably makes me and the others uncomfortable—he's just weird, I guess!

Read Twain's *Literary Offenses of Fenimore Cooper.*

Writing in dayroom! Watching "That's Incredible," about drag racing the human torpedo. The stereo is playing and two TVs.

A man is being dragged on his belly at 165 mph—prostrate on titanium plates, sparks all over. That's incredible—crazy, I guess.

29 December 1981, writing on the thirtieth

Nothing much happened. Very nice sunny day. Walked to the post office and bought desert commemorative stamps for Patty Grippen. She walked with me a lot. Boy, she sure is cute, smiles so often, and is affable to all but tries to maintain her distance. It seems she is trying to be secure by having many male friends. I never see her talking to any girls or eating with any. I walked to Walker Center and spent the rest of my money. I reread *Fenimore Cooper's Literary Offenses* while at the dayroom at night. Seven people were dancing to a heavy-beating, siren-voiced disco song. The song lasted about five minutes, and every time it would end, they would just replay it and listen to it over and over about ten times while I was there.

30 December 1981

It's 7:50 A.M., a beautiful cold clear morning at Fort Leonard Wood. About twenty-two degrees, but there is wind.

Poetic justice today: A Private Best was told to carry road vests back to the barrack while we were in formation. The reason; last night he cut across four lines to get in the first line to eat. Snyder and I let the sergeant in on the secret, and he paid for it! There was honor in *our* ranks among me, Snyder, and his roomie. My roomies are Hanks, Lackadoo, Thomas, Penn, Newkirk, and Marks.

Yesterday Private Lackadoo was acquitted for having alcohol by First Sergeant Williams. The witness said he didn't see him do it (possess a bottle). Lackadoo does know who did it but would not tell.

At 9:00 A.M. I just finished a bit of detail for Sergeant Williams. We had to get on our hands and scrape the paint off the floor below the baseboards. They were painted yesterday in the three offices of the cadre of C-4-3. Depending on the individual office, the color varied: blue in the staff sergeant's office, black in the first sergeant's office, and the most hideous red scarlet in the captain's office. All the walls were the same blue-gray—dull. I would probably not have even mentioned this small detail this morning except for the strangeness of the color contrasts in the office.

His office is about twelve feet by twelve feet. Doors were on a side wall and in the corner of the back wall. He or rather his desk faces two windows that are four by four together in front of his desk. These are covered by thin curtains that are also the same hideous red. When the sun shines through these, the room is cast aglow in the same hideous red. Blood red. Imagine the scene, two scarlet doors, a scarlet baseboard, and faded blue-gray walls, all of this bathed in that evil red glow. Prior to painting the doors and baseboards red, they were a deep bright yellow. I don't know if the curtains had changed or not, but he must be strange or color-blind.

At 9:10 I'm thinking about army life while talking to Pvt. Marilyn Prior. We are the first generation in history to have the large advantage of joining the army in peacetime, going traveling around the world, and learning a technical trade. Never before has the army been so technically advanced.

I asked Prior why she joined. She said she needed a change in her life; she had been working so hard and wanted out. She wanted something that was a challenge—and she emphatically said, "This army is a challenge." She has heard about the army from her boyfriend, from her sister—and both told her not to join, but she had to find out for herself.

She says she does not like it all that much and that sometimes it is difficult for her.

It's 11:20 and finally we have received all of our tardy letters and packages. Yesterday we received two letters for twenty-five people. Today we received four. Something was wrong. People of our holdover company began complaining to First Sergeant Williams that they hadn't received their packages and letters sent to them from home. The first sergeant took about six people over to the mail room, and we found out that all of our packages had been packed away in a corner. If anyone out there is reading this and has a son or daughter away, yes, they are ecstatic to receive packages and share them. They make everyone happy—yes, this is contraband. The result of receiving all of these packages—about ten of them—was a sweet pig-out. We had boxes of chocolates (Whitman's Sampler), bags of cashews (three pounds), three-pound bags of Kisses, three-pound bags of chocolate stars, several boxes of cracked chocolate chip cookies, a big box of popcorn, boxes of fudge and brownies, apple turnovers, a bag of venison sausage. There was at least twenty-five pounds of goodies. The result of all of this hitting a bunch of unbusy recruits was a feast. People were laughing and stuffing their mouths; they were running from box to box and just making complete pigs of themselves. It was great.

Lunch was soon after that, and I went simply to stay on a regular meal plan. I wasn't hungry.

Once again I chose the wrong line in the formation. It is anybody's guess which line in the formation goes to eat first. Since we get a new CQ every day, trying to bet on which line will go first is a real shell game. I am losing too much money! I shall stick with horses. You will encounter a few people in the army who won't take a shower for love or money. They won't shower until ordered to by a DI. Yes, they *do* stink.

A fad or cliché used very often here by the DI's is, "You're wrong." They use this often in the sense of "if you do not have a hat or coat on, *you're wrong*" for emphasis and out of habit. "If you haven't shaved, you're wrong"—and so on in that manner.

The recruits pick up this line and satirize it to all extremes and outcomes:

If you're white, *you're wrong!*
If you're black, *you're wrong!*
If you're in the doorway, *you're wrong!*
If you're Puerto Rican, *you're wrong!*
If your nose is running, *you're wrong!*
If your ass stinks, *you're wrong!*
If you're asleep, while sleeping *you're wrong!*
If you shit, *you're wrong!*
If you are wrong, *you're wrong!*
The whole *fact* of the matter is: you are wrong!

A new DI for us today—his name is Sergeant Jones. I asked him what his name was (so I could write in this journal), and he yelled, "Sergeant Jones, why?"

"Oh," I remarked, "I just wanted to know, Sergeant!"

But that didn't satisfy him. "Why duh ya want to know my name, Pravate Stanley?

"Oh, Sergeant, I just want to know who *I* am *working* for."

That built his ego up enough, I think, and he pried no more. (I really wanted to know his name for the purpose of using it here.)

To me, Sergeant Jones portrays the classic good-old-boy sergeant with a lot of latent seriousness. He is about six-one, slightly filled out, sarcastic as any person I have seen, and loves to chew his tobacco in a big wad in his left cheek. He will stand on the edge of a step and after speaking three or four sentences will spit a wad out. His face is plump but has no double chin. He says, "Dis groun' flo is for duh females and duh females ownly. No males are to be talkin' here to dem. If y'all git caught, yo be working fo me tonight! Fo those in Alpha, Bravo, Charlie, and Delta companies, I make them scrub duh halls with yo toothbrush" (spit). "Also, if yo all git drunk and make me come gitcha, the same benefits will be giben to yo all" (spit). "Y'all know what tahm we have formation agin, don't ya?"

Battalion: "Yes, Sergeant."

"What tahm?"

Battalion: "Sixteen hundred, Sergeant!"

"Are dere any questions in those little brains of yos?"

"No questions!"

"Are yo all sure yo hab no questions for me?"

"Yes, Sergeant."

"Yo all not askin' questions so I can release yo on pass sooner, arn't yo all?" (Spit.)

"Yes, Sergeant."

"Okay—Battalion! [Battalion to attention.] Fall out!"

While I was at lunch today, waiting to enter mess hall, another DI walked by our platoon and noticed one of our recruits with a bandage around his head and covering his ear. The sarge asked, "What happened, son?" (DIs always refer to recruits as "Son.")

"A stick poked me in the ear, Sergeant."

"Ah don't believe that, son. Y'all must be going out with them heifers. Y'all better watch out fo them. Heifers, they kiss hard."

"Yes, Sergeant."

I ate lunch yesterday, of course, and attempted to sit at a table in the back of the mess hall. I have sat there any number of times without incident, but today there was a sergeant sitting at about halfway down the length of the table. These tables are not short, and I would say he was about ten feet away from the end (which was where I was going to sit). Normally, most sergeants sit in the front of the mess hall at a few various tables, but tradition really doesn't matter to sergeants.

He snapped at me, "Do you know me?"

I replied, "No, Sergeant."

"Then you'd better sit somewhere else!"

"Yes, Sergeant!" No big deal. If he wants me to move, I will. No sweat. I moved two feet to another table. I hope he got heartburn.

There is one thing for sure that I've noticed here—since I have been here, I have not had one congenial conversation with a sergeant and I have been here a month.

Money, it seems that everyone wants to borrow a quarter here, a quarter there, if you are near them. I sit in front of the Coke machine on a gray love seat that has a modest amount of love. It is vinyl-cushioned. It could have been stolen from a bus, I suppose. It is like everything else in the army—not pretty, but functional. The Coke machine and seat are off in a small room adjacent to our dayroom. It is the quietest and most comfortable spot in our battalion area. We have no chairs or tables in our barrack. *Never* sit and write in front of a Coke machine. If you do, you are setting yourself up for an assault by panhandlers. Almost everyone that comes to the candy machine will look at the price, dig in their pockets, finger their money a bit in their palm, and stop. They will then look at you and say, "Do you have a nickel?" or, "Do you have a dime?" I had two nickels and a dime and also a quarter. My last! Thank God my money is gone. Now I am innocent.

I found out today that my roommate Thomas lent out $200—I don't know when. I advised him as a friend that unless he plans to lose money, it is not a good idea to lend money. I shared my experience with him: When I lent $175 to a "friend" while I was in college, my friend, of course, disappeared. I hope Thomas won't have the same experience. Tomorrow is payday and all of this will end for at least a week. We get paid at the end of the month. Once a month.

Talked to Thomas today. Says he joined the army to get experience in electronic engineering. He is in the National Guard. Says he worked at Kodak on the assembly line and needed experience in the electronic field to move up into better jobs at Kodak. Says his boss advised him to join the army for experience.

Went to the gymnasium today—took a sauna for one hour and worked up a terrific sweat. Saunas are great, because one can work up a sweat without working. It is cold—thirty degrees—a windy, clear day, sunny. I saw a red-tailed hawk banking and hunting over the battalion area—a poor place to hunt for food. It is ironic to see such a free sight in such an unfree place as Fort Leonard Wood.

31 December 1981, Thursday, New Year's Eve

I have been here thirty-one days, so now I have three years and eleven months to go.

Litten came down the hall to wake us up and said with his drawl, "Who is steel in baid? I wont to throw eem on hees haid." He would talk real slow and sound like a dragon, would walk real slow into each room and scan to see if anyone was in bed.

Our first formation was at 6:15 A.M. today. Pitch darkness and stars outside. I remembered that Sergeant Litten had said our breakfast formation would be at 6:40.

I went to his office and asked.

Bryan: "Sergeant Jones?"

"Yes!"

"Sergeant Jones, when is formation?"

"Shortly!"

"Sergeant Jones, I thought you said our breakfast formation was at 6:40 A.M."

"Yes, I did. But we are having another one now for policing the area."

"Sergeant, it's dark out. We can't see. You mean we're still going to have policing?"

"Yes, the ahmy does stupid sheet like att."

All during this short dialogue, he, Jones, was just halflaying in his chair, leg over the arm and a big round cheek full of tobacco, and of course his sheet-eating grin.

Another drizzly Missouri day. This weather is meeting and exceeding my abysmal expectations of here. It is great; it is magnificent—such a drool!

Got a talk to our Christmas rejects from First Sergeant Williams today. Pointed out that our picnic will be over here at Fort Wood this Saturday morning. Said our vacation will be over.

Said we've had things pretty good here. Is letting us have another party tonight—this is New Year's. Said and reaffirmed that C-4-3 is the best company on the post—not only DIs, but trainees as well. Said he has been satisfied with our behavior over Christmas. Said there is a lot of maturity in our group that stayed over for Christmas, not only in age but in emotional maturity. Said he is going to need our help for the incoming personnel this weekend, said we have had a good experience and that we should share it with the trainees that are coming in. He said that the other trainees were here for only two weeks, they were ridden hard, and they will be ridden hard along with us starting Saturday. He says we have seen that the army can be much better than during total control and Basic, but that other recruits haven't seen this and this can make the trainees feel hopeless. Williams also says he will be riding us hard for doing anything wrong, anything. He says he will be on us like "white on rice." He says the DIs are just doing their jobs and they are coming back *ready,* ready and rested. We shall see, I believe.

I still must add some more statements of admiration for First Sergeant Williams. He is very knowledgeable and can see subtleties and realize the importance of them. He is concerned about us.

Berraras is very violent today —he has been drinking and trying to pick fights all day. Tonight in the barrack he called several blacks niggers and would have tried to fight them all. He would have gotten hurt. With much talking and separating we kept a fight from breaking out. All day he just wanted to fight with anybody because they were "prejudiced." I can see why the army does not want anyone to drink. There are some who cannot handle it. Mike Snyder reminded me yesterday that Berraras had a chip on his shoulder, but I said I hadn't noticed it! Mike reaffirmed his observation to me, and I had told him, "You may be right." He was—Berraras continued his obstinance at the party. I was smoking at a card table about four feet from where he was sitting and blowing smoke away from him, and he started calling me an asshole for smoking and saying I was provoking *him.* Another example: Snyder was just watching the TV (a football game), standing there and chewing on a sucker. Now Snyder is a modest guy—even did some preaching at his Baptist church—smiles a lot, is friendly, and shares his feelings freely. A great guy, he never bothers anyone.

Well, Berraras just walked right up to him and stared in his eyes from two inches away. Snyder just turned his head a bit and the sucker stick he was chewing on just touched Berraras's nose. They just stood and stared at each other. They just stood there. A face-off. Neither moved; neither flinched—for about two minutes. Finally, Berraras walked away grumbling, and Snyder walked away laughing. Berraras is a Hispanic, a dark, chunky person.

Tonight at the party, our New Year's party, plenty of food, plenty of soda, coffee, and punch, and plenty of people. I think we spent $120 total for food and drink. People were playing charades and mostly imitated movies and TV—"Hang 'Em High," "Love Boat," "Welcome Back, Kotter." This is the first time I have ever *watched* charades, as I was afraid to participate. I was afraid to get up, I guess.

We continue to play Spades—Wheecks, Patty, Snyder, and I. We alternate teams. It is fun but gets repetitive. There are TVs going at once. And the programming is varied, from boxing, to pro football, and "Magnum PI." And also people are dancing, always dancing, at least that's what they call it. Bobbing here and rubbing there—clapping and singing. I like to see people have fun, but they could change the music. They have three disco songs, and they keep playing them over and over and over again. Big bass drums—bomb bomb bomb bomb bomy bom bom bomp bom bomb bomb bomp bomp bomp bomb bomp. We have a very fine FM station that plays FM stereo rock and roll, but they claim it is no good for dancing. They also say that *I* am the only one who wants the radio or at least the only one who asks for it. So I took a poll. About eight people were dancing, and about thirty were sitting down. A random sample from all of my friends revealed that we were unanimously in favor of the stereo FM being on. But did they play the stereo? Of course not.

I talked with Carlos Hemrick a few minutes ago. It is about 10:30 P.M. He is about the only one who really has an idea that I am writing this story. He asked me if I had written anything about him yet, and I replied, "No." I said, "Up until now I have been recording mostly crazy stuff, and you are not very crazy." He then said, "Oh, I'm too normal, huh?" I said, "Yes," but then thought I hadn't been objective enough in writing and providing a true glimpse of the army. I have not been complete in my coverage. I should be! I excused myself, with the fact of the matter being that actually there is a tremendous amount of subtle activity going on here and I only have about enough time to write about three or four handwritten pages a day—well, maybe five or six—but it is very difficult for me as a novice writer to separate the wheat from the chaff.

Hemrick is a Vietnamese-American, born of a Vietnamese woman by an American serviceman. He came to this country at age one. His mother is still in Vietnam. His father's sister raised him in Seattle. A quiet person. And takes an interest in people and appreciates a good joke.

Before supper tonight, I went up to my room. Lackadoo was lying on his bed asking if I was Joe (Hawks). Lackadoo did not look to be in a very ambitious frame of mind—he looked slightly inebriated.

"Joe? Joe, is that you?"

"No, this is not Joe. This is Bryan."

"Joe? Are we in the barrack yet? Joe? Where are we?"

"I'm not Joe, and we are in the barrack."

He passes out. I go to dinner.

I find Hawks at dinner (Baby-faced Hawks)—he is slightly inebriated, too. I mention that Lackadoo is slightly toxic and talking to himself. Hawks goes on to tell me how Lucky drank a quart of rum. I was astounded by the knowledge of this and began to worry a bit. I knew that Lucky would pass out. I had known people who had drunk as much in Stevens Point and never lived to tell about it. After dinner I hurried back to the barrack to find that Lackadoo was up and around. He was staggering a bit and bleary-eyed, but he greeted me and I him.

"Lackadoo, I am glad to see that you are up and around!"

"Oh, yeah—I'm feeling pretty good."

"Thought you would be passed out."

That was about the end of our discussion, but I saw him later and opened it up again.

"Lackadoo, I'm impressed that you can drink a quart of rum and still are walking."

"What? Who told you I drank a quart of rum? I had about this much [shows fingers] and left the rest. Hawks is full of shit! You see? This is how rumors get started!" And on and on.

At the present moment of this writing, Hawks is sitting across from me at a table at Walker and *he* insists Lackadoo is full of shit! Well, anyway, someone is full of shit!

Lackadoo continued his fraternizing last night and picked up that Johnson girl—she is big, about six-one, blond, bucktoothed, has two kids, is not married. She has been looking for someone ever since she got here at the Christmas station. He danced with her last night and ended up taking her outside. She wanted to give him a blow job, but he didn't want to take his cock out of his pants.

1 January 1982

Got up this morning and felt rather chipper, even though I didn't sleep very well. I had an urge to imitate Sergeant Litten this morning—a more satirical urge, I believe, than a respectful one. And so without thinking much I dropped my lower jaw and with a fair Wisconsin attempt at a southern (Missouri) drawl let out, "If'n y'all ain't aouta yo baids, I'm bustin' yo haids." I was unaware that a few trainees were still sleeping (lazing in bed) across the hall, among them Hemrick, Berraras, and Martins. They immediately panicked, tore and tossed at the sheets, and scrambled to get to their morning duties. This, I believe, was a fair test of my voice imitating ability, though admittedly a bit crude. Hee-hee!

2 January 1982, Saturday

Weather windy, light rain, temperature thirty-five degrees, time 7:30 A.M.

At 9:00 A.M. It's still pouring rain; it is getting wonderfully more dismal. Writing in Charlie Company barrack. It is good to be back, even though the Alpha barrack got to feel like home. It will be good to get back into the more disciplined crowd of C Company.

I feel good—a change of some sort usually does feel good after it occurs. Going through the process is mysterious and anxious and work, but afterward it is good.

Sergeant Gibson gave us holdovers a lecture in his southern drawl with his snappy style. He made himself clear. We are back under total control. No drinking sodas, no fraternizing or even talking with women, no smoking when we want, no kicking back and laying out when we want. The party is over. I will add that the two-week break did feel good, and even though we didn't get all of the free time they promised, we did get enough. I met many new people from other platoons and companies and made friends and also got to know my previous acquaintances better. I feel the people that stayed here over Christmas will be at an advantage (like Sergeant Williams said) over the returning trainees from Exodus. We will have more and better friends here, we will have seen the everyday life activities of Fort Wood (PX, bowling, gym, Walker Rec Center, library, NCO club, post office, and restaurant, to name a few). Another is that we got to see the DIs (a few) in a *more* relaxed mood and can realize much better now that they are actually just doing their job.

Wow—it is still pouring outside. I am looking out the window and the athletic field looks like a muddy brown river delta. Huge silt-laden rivers of runoff all heading to one end. Many other smaller tributaries, too. Heard news from Wisconsin. It's been below zero there for weeks.

I taped up all of my Christmas mail on the interior of my locker door. It is a bit old now (but it still gives me good feelings to look at it). Cards from Boland, Racette, Skrivseth, Heidi and Linda, Fleishman, Warren Johnsons, Burkharts, Barb Jones-Cass. And postcards from Marrakech, Andorra, Iowa, and Missouri—they really add some flavor.

The barracks are pretty quiet right now.

I ate breakfast this morning—the last dinner I will have at the AIT (Advanced Individual Training) cafeteria. When we finish eating, we must carry our trays through a door, in a line of course, dump our paper, dump our garbage, dump our silverware, put cups in a rack, stack our plates, stack our bowls, then put our trays on a rack and then head for the door.

Well, I was not sure what to do with my cereal bowl this morning, and so I placed my bowl on a plate stack.

The woman washing plates then said, "Son, you stack you bowl on the bowls." (They always call you Son.)

I turned around, did this, and then said, "Yes, Mom."

I began walking out and then heard one of the other dishwashers say loudly in a laughing voice, "Yes, Mom. Did you hear him? He said, 'Yes, Mom.'"

I failed to hear the rest of the dialogue, but Snyder was behind me in the line and said it concluded like this: The woman said, "I don't care if he calls me Mammy Map, as long as he stacks the bowls with the bowls."

This morning while we were waiting for breakfast in the

Alpha barrack, a few of us—Wood, Snyder, Lively, and myself—were just sitting in the hall and conversing. The dialogue drifted to the topic of Peterson, a person who has in many ways made people take notice of him. Not only by the way he smells and plays cards and condescends to everyone, but he does not sleep on sheets and doesn't change his clothes.

The first day in Alpha barrack, Peterson walked into Wood's room with a very serious countenance and said to Wood, "I've got a problem." Wood, being a man of fair integrity, thought, and having a companionable temper, decided to help out and listen.

"Okay, what's the matter? Is there anything I can do to help?"

"It's girls; they are my problem. They love me too much, and I love them. I can't control them. They can't stay away from me."

From then on, Wood said it was a monologue and Wood was dumbfounded, as the rest of us were. How could a stinky ego-head big-nose like this have such a problem? One half of his eyes is half asleep most of the time and when asleep is nearly closed and twitches about every five seconds.

Wood said: "Sure, Mack. I sympathize with you! But have you tried to slap them up, kick them in the butt, and starve them?"

Bronas told me the other day that Peterson offered to protect the girls. He also went to every girl to try to pick them up. He started from the top and proceeded to the bottom, where he is now with a dejected handy wallflower. Birds of a feather.

Just got back from lunch, marched over with rejects of Charlie Company—in our big, bulky field jackets and raincoats, we looked and felt like stuffed toads. Rain is still pouring.

It was good to get back in a formation at C Company and to practice our facing right-left about—it was good to yell, "Charlie!" after "three-four."

No more twenty-minute showers now that we are on total control—just very short two- or three-minute showers.

After lunch just practiced reporting to an officer, treatment for shock, treatment of a wound, treatment of a broken leg, and application of a splint.

When I am reporting to an officer, my salute quavers and I am tense. I am getting better with practice—Sergeant Hensen says, "Very good, with some minor adjustments."

We are also practicing those minor medical treatments. I often volunteer to be the victim, because it gives me a chance to lie down on the job and relax.

Orfino is still complaining about getting out of the army. He keeps maintaining the army is "not for everyone." He and Snyder argued a bit about it today! Orfino has also quit school (high school). He says he wants to go to the Berkeley music school in Boston—Orfino does play the harmonica well.

Snyder says, "Why quit? You will just quit something else anyway."

"No!" says Orfino. "I only quit high school."

Snyder remarked (because Orfino wants to be a mechanic also), "Well, I wouldn't want you to work on my car. You might have a problem and quit on the job."

"Oh, fuck you, man. I wanted out of the army two days after I was in."

I actually want Orfino out myself, because everyone else is being affected by his negative attitude.

A little while—a few minutes ago—Hensen spoke to Orfino to tell him even if all of his papers were filled out today, he would still be here three more weeks. Hensen told Orfino that if he tries hard to like it for another two weeks and he still doesn't like it, then Hensen would do everything he could to get Orfino out. Upon his hearing this, Orfino's attitude changed immediately. Upon coming back to us and informing us about what Hensen said, Orfino felt good about it and even said, "Well, after two weeks, I will only have five weeks left and maybe I will change my mind then."

It's 7:00 P.M. and we have personal time until 8:10—that amounts to about one hour and ten minutes from now. We have had personal time since 6:30 P.M. Tonight we drilled our drills for a while, marching about-face, right-face, left-face, left and right flank, change direction, and practicing putting on the gas mask.

The gas mask is made out of rubber and has a thin vinyl hood that pulls over the head and neck. It has two baseball-size windows for eyes and a round filter where the mouth should be—that is the air intake filter; it is also about baseball size. On the cheeks of the mask are two exhale valves, and they are about as big as golf balls. Altogether gas masks give you the countenance of a green frog crossed with a fish—I hear they are functional; I hope they are.

We have a small pouch that resembles a handbag. It fits snug on our left sides under the arm with a shoulder strap and chest strap. The opening is in the front. On our hearing the order "Gas," the flap is popped open with the left hand and the mask, which is set in the snug pouch with the face out, is then grabbed with the right hand. In as rapid a manner as possible, the straps on the mask, which is practically identical to a catcher's or umpire's mask, are pulled back on each side, one side in each hand. The chin of your face is then put in the chin opening, and the straps are then pulled over your head. They (the masks) are not pretty. After getting the mask on, we run our fingers along the head-mask interface to make sure the rubber seal is formed correctly to prevent air leakage. We then plug the exhale valve and blow to clear the mask of any foreign substances. We then cover the inhale valves and suck in air; this is done to make sure the mask is sealed. If the mask is sealed around your head, it will collapse somewhat when you suck in. After that, we raise our arms and rapidly wave our forearms in a flexing-the-biceps fashion. With our elbows parallel to our shoulders and our fists near our heads, we do this three times and yell, "Gas, gas, gas, this is the alert!" We then finish pulling the hoods over our heads. (The mask is not comfortable but does not feel all that different from a scuba mask with aqualung attachment.)

We must do all of this in fifteen seconds flat or faster. We will be tested, too. How else? By tear gas. Upon hearing the signal, "All clear," we can remove our masks. We are told sometimes to remove the mask by their saying, "Take off the mask." If we take the mask off on this order, we do push-ups. We are told that sometimes when a company is on a long march, a gassing is done on the company. If one is not fast, he will be gassed. Often, too, the DI will say, "Take off the mask." Some will and, of course, will be gassed. Tsk, tsk!

3 January 1982, Sunday

It's 7:00 A.M., a gray, foggy morning. It is not light out yet but is getting close to it. The distance of one hundred yards is all blue, and beyond that all that is observable are the points of light from the streetlights and the cone of illumination that passes down to the mists below them. The light from the sodium vapor lights is a lurid glow and is more easily seen than the white streetlights. The forms surrounding the field look just like they have a false front. They simply look like a wall, with no depth.

McCubbin, Vieths, and Reynolds are back, which almost completes our room—except for Pearson.

Vieths' mother and father broke up after twenty-eight years of marriage—so he didn't have a real happy vacation, and it is still affecting him now. He said he would give his left hand to have all of this Basic over with and have this be the end of February.

Reynolds's wife is two months pregnant and has an infection in her Fallopian tubes. She is going to have an operation. He wanted to stay home but couldn't because the Red Cross couldn't get in contact with anyone at the fort. So he had to come back to avoid being AWOL. Sergeant Hensen was very perturbed by the fact that the Red Cross couldn't contact the fort and said Reynolds should be home.

It's 7:15 now. Two days ago I started thinking seriously of going to Airborne school after AIT (Advanced Individual Training) and also going to OCS after AIT. I guess I will admit to a little fascination with the exciting Airborne training—it just seems so exciting to go Airborne, mostly because of the money. OCS is starting to sound more attractive also, but I am not sure what division of the army I would have to go to. I hear the demand for officers now is in artillery and infantry. I don't know if I could go for one of those and also maintain my career field in language and intelligence.

There is much more marching and tumultuous marching songs going on now since everyone is returning from Exodus. On these warm, foggy, mornings the low voices of a platoon carry for so far—first you will hear the low, melodious voice of the DI, followed by an amplified call by the platoon. I can't help but admit to a good feeling when I hear a DI and platoon singing on key. One of these platoons just walked under my window a couple of minutes ago. I had to stop my writing and watch them, not only to watch, but to listen. Also, they seem so organized but flexible; they are soldiers. I can tell by the way they march and sing that they enjoy it.

It's 9:00 A.M. now. I went to church with Hanks, Kawalski, and Osterholm, to Catholic mass, and just got back. Father Maj said mass—he emigrated from Poland in 1971. He talked about Poland a bit today. He said his sister helped organize

Solidarity and his nephew is very active in it. He has not heard from either of them since martial law has been in effect—that is about one month now. We are all concerned about Poland—many Polish people here in this country have relatives there.

The priest also talked about us soldiers and said how our mission is one of peace—we are here to maintain peace and freedom in the world.

Things are hot in the world right now—Latin America, Eastern Europe (Poland), the Middle East, South Africa, and Southeast Asia. Most of it is leftist (communist) terrorists in an action to overthrow governments and establish dictatorship proxies.

It's 10:20 A.M. Just got through talking to Sergeant Hensen. I had a few questions about getting military flights to Europe (a hop), Airborne training, OCS.

I was pleasantly surprised to find out that I could get a flight from the US Air Force bases to European air force bases. The most common flights are from New Jersey (McGuire AFB) and South and North Carolina AFBs. From there the most common flights are to England and Frankfort, West Germany. Total cost is $10. That is an incredible rate, since the commercial rate is about $700 or $800 roundtrip. As soon as I can, I am going to Europe.

I also talked to Hensen about Airborne—he says I should wait until I get to my AIR locale and talk with the Airborne rep. My MOS is a voice-interceptor intelligence, and my language specialty will be German. He says that most intelligence is gathered from the soldiers up front, but the army needs intelligence people there to intercept radio signals and translate them and also interpret and collect intelligence from footsoldiers. Most army broadcasts are only good from one to three miles, so interception must be close to the front lines. So that looks good.

I talked a bit with Private Calloway, and he discussed Ranger school—Airborne Ranger. He said they are trained to climb cliffs, sneak through swamps, and much of everything else behind enemy lines. That sounds very attractive and exciting to me. My only worry is my foot. It gives me pain sometimes when I run. It's my left foot. I dropped a tree on it once when I was logging near the peaceful evergreen shores of Lake Dubay.

It's 11:00 A.M. Just some more notes. Saw Alan Berella today at breakfast. Good ole quiet Al! He got married over Christmas to Debbie. He was originally going to be married in a few months, but since his time would be limited, he opted for it now. Surprise, surprise, surprise! He suggested we start a junta.

We know we are back in Basic Training and total control—last night we went to chow at the AIR chow hall and Sergeant Florez put us in the front leaning rest position and made us jack out eight four-count push-ups. He then made us hold the FLR for five minutes and said, "It's over, folks; it's all over," and proceeded with his lecture about it.

Today when I got up to third floor of C Company in the Second Platoon area, I found Second platoon in the FLR in the hallway. Sergeant Florez once again, and he was giving them hell: "No, I'm not bo-shittin'. Are you all going to follow my details now?" "Yes, Sergeant Florez!"

I talked to Lackadoo at chow tonight. He said Florez was not in an agreeable mood today. He put twenty-one people in the FLR for thirty minutes. Lackadoo said he went to Florez to practice the "reporting to an officer." Lackadoo checked his uniform and his buttons. Florez said, "You didn't check your buttons good enough—git down in FLR." Lackadoo did ten for thirty minutes. He said the first fifteen minutes weren't bad, but after that he started to lurch and shake. Yes, Florez, we know we are back.

We also know we are back to our mess hall. We are back to eating our dinner like wolves (to save time and space). We must be out before other companies come in.

K. Thames (Second Platoon) came into my room after dinner and talked a bit. He started talking a bit about Sergeant Williams—the redheaded, fiery DI. He said Williams yelled this statement in the chow hall: "Basic doesn't start until tomorrow! You ain't seen anything yet—I'm gonna fuck you all up—yeah! I'm gonna fuck you up! Yeah, I'm gonna kick yo asses!"

While cleaning the Charlie barrack a couple of days ago, Williams came in and yelled, "You boys just wait until January 4! I'm gonna be just like a football player! I'm gonna be the offense and you're gonna be the defense, only I'm gonna shut you out 68–0."

4 January 1982

Today was cold, below freezing. Sergeant Monday said tomorrow would be fifty degrees Fahrenheit. Notice the change of weather between days.

After our morning class we even filed outside of the first-aid barrack to drill awhile for lunch. I was flawless—I did every command correctly. We lined up for chow among the sickly-looking sycamores with their white, scabby bark, twisted trunks, and empty crowns.

While we were in line, Sergeant Reid checked us for rank insignias and I mistook a first lone bar for a captain—ten push-ups. I then named a colonel's leaves but was charged ten more push-ups for missing the lone. While in line, I saw many others in the FLR. We ate outside the first-aid station between barracks, and that was more like a wind tunnel than anything else. A little fresh air is always nice during a picnic. Al Berella said that this was paradise compared to the Russians' training. A lot of blanket parties there.

There was a gravestone among a few oak trees that said:

RIP
NO
FIRST
AID

When one is wearing the pile cap, it resembles a cow pie
on someone's head when rolled up. When pulled down
around the neck, it resembles a beanbag rolled over the head
with a hole cut out.

5 January 1982, Tuesday

It's 5:20 A.M. and I just got in about ten minutes ago—
went out for PT at 5:00 A.M. after getting up at 4:30.

Clear skies; kept drilling us today. We were still yelling
"More PT, Drill Sergeant, more PT—make it hurt! Drill Sergeant, make it hurt!"

Can you imagine a crowd of grown men pleading for this
type of masochistic treatment? We often said, "Fuck you, Drill
Sergeant, fuck you! Go to hell, Drill Sergeant, go to hell!"

We found out tonight how much the army trusts us. They
confiscated all of our civilian clothes to act as a precautionary
measure against our going AWOL—they said we could keep
them before.

Weather today warm, forty-five degrees Fahrenheit.

Went to NBC (Nuclear, Biological, Chemical warfare).
Some notes in red notebook. Ate lunch outside once again
(beef stew, bread, mashed potatoes, lemon pie, coffee, and
milk). Food delivered in deuce and a half.

Sergeant Gibson gave us a pep talk today—said C-4-3 was
the best company on the post, said he was the best senior drill
sergeant, said his DIs were the best. Said most records are
C-4-3. Said A, B, D, E companies say, "God, I'm glad I'm not

in Charlie Company." Said he isn't going to take a half sound-off from us.

Sergeant Hensen related to us while we were in formation how a guy who was sort of an asshole in training (BT) and mistreated people finally got graduated from Basic. He went to AIT in Supply at Fort Leonard Wood and finished with that. Then where did he end up at for his first duty station? That's right, back at Alpha Company.

6 January 1982

Morning forty-five degrees, evening fifteen degrees. Quite a drop in temperature, but was not the main event of the day.

We got the gas chamber—and I can say it was one of the worst experiences of my life. I thought I was going to die. It felt like someone was grabbing my lungs and pulling them out of my body—arg-g-g-gh!

I had been screwing up all day, put my protective coat on when I wasn't instructed. I lined up on the wrong side of the platoon. I only got three out of five guesses on the decon test. After having just listened to instructions, I am rather spacy today. Girls, no doubt.

All day long we were being briefed on gas attacks, biological attacks, nuclear attacks, etc. We mostly concentrated on the protective mask and related maintenance. But I really didn't relate to it all that well, because I had never had a teargas sample—I also didn't understand the effectiveness of the protective mask when one puts it on under normal air levels. One breathes with it well whether it is on or off. Actually, the breathing is a little bit harder with the mask on. But when one goes in the gas chamber with the mask on, he doesn't notice any difference in his breathing when he walks in from the pure outside air. But, oh, when you remove that mask, you realize it is effective: the gas and the mask. Some people dropped their masks or pile caps in the gas chamber and had to go in and get them.

Our gas chamber was a short hike (half a mile) from the NBC center. It was in an oak forest area, almost parklike. It was a rather innocent looking building about the same shape and size as a one-car garage. It was painted camouflage. The only thing that distinguished it from any other small utility building on base was the fact that it had three signs on it that said GAS CHAMBER and a curious white smoke seeping out the doors that made it look tavernlike.

Our entire platoon was formed into a line, 225 bodies; procession groups of 15 would file into the building, line up along the wall, breathe through their masks for a while, and on command take off the masks. After each man had removed his mask, we filed out at a fast walk. The gas was being emitted from a smoldering coffee can. I couldn't tell if it was a fire or a compound in water.

I was about halfway down the line and next to Snyder. We all had our protective masks on, so it looked like we were going to do auditions for a science fiction movie—*Green Men from Mars*. We would notice these groups enter and then exit in about one minute on the opposite side of the garage, most stooped over and coughing. Occasionally one would burst out of the entrance coughing and gagging like in a convulsion. They looked like they were uncomfortable, but I couldn't understand how uncomfortable they were.

Before going in, I remarked to Snyder that I was going to take some deep breaths so that I would have a better experience to write about, and he said he was going to hold his breath. I asked a sergeant if it would be harmful to breathe deeply, and he said it could be if the gas gets trapped in your lungs. That made me think a bit about if I wanted to breathe deeply.

Well, it was finally my turn to enter and I was the last one to be accepted in this particular group. We filed in past white smoke and noticed that it was very smoky inside—a man was in an enclosed glass booth, and he would give the order to remove our masks. The whole thing had an evil, conjuring look to it: dusky lighting, the smoke billowing up out of the coffee can, and all of us dressed like alien spacemen.

We were ordered to remove our masks, and so we did; I contemplated taking a deep breath from the mask before taking it off, for I was breathing normally and without any effects whatsoever. The whole situation seemed like a fraud. I removed my mask and just felt a slight stinging sensation in my eyes and on my forehead. This was a cinch! A piece of cake. This didn't affect me! I'm superhuman. I became smug. I started to breathe, and my mouth and throat became an inferno. It was time to move and move fast. I couldn't believe the intensity of it. I felt like I was swallowing fire. My first involuntary reaction was to cough this pestilence out, but that was a mistake, too. I should have held onto the burning, which went simply to my larynx, for after I coughed up all of my air, I promptly sucked in a lungful now in the hope of getting some sweet air in my lungs. But there was none to be had. I was in a complete state of disbelief at the conflagration burning me up from the inside. I made it out the door finally, but that did not help much at first. At that moment, I knew where every square inch of my lungs were and where all the bronchi there turned. If someone were to ask me where my air passages were, I could have traced them on my chest for him. I started bellowing like a cow: uh-unnn, uh-unnn, uh-unnn. Snot was streaming out of my nose, down my chin, and on my coat: I didn't care. Tears were streaming down from my burning eyes. From the sights I saw of others, I could imagine well the dreariness, the bloodiness and pain of mine. But more than that was hack after raw hack that racked my chest and diaphragm. The flames were reaching from my gut to my mouth. Flames, flames, there had to be flames with such anguish. There had to be. I stumbled and stooped across the road to the other asphyxiants, bellowing and bellowing all the way. I felt the urge to vomit but somehow didn't. From the belching I was doing, I knew that I had swallowed some of that contaminant. I couldn't believe it could hurt so bad. I don't know why they call it teargas; they could probably come up with a better name for it. More descriptive, at least.

Gradually the pain went away and I was able to breathe and talk again. Many of us *were* in pain. Some were laughing at others coming out, but those must have been some who didn't take the full poison. I was happy to be alive; I thought I was going to die. I gained a profound respect at that moment for my protective mask and a more profound respect for the harsh gas chemical that we might be subjected to. I also felt a very deep pang for the 1,300,000 men who were gassed to death in World War I.

PT tonight in the barrack because it is too cold outside. Sergeant Florez had the B squad up on the third floor—every time we would regain to attention, he wanted us to yell our company motto: "Duty, honor, country, drive on, DS, drive on!" He wanted us to break the windows with our shouting. He says he has broken three windows in two years as a DI. One time, two times, three times in we would try, but we failed this time. Everyone would plug their ears as we blew (including Florez). We had about forty of us in the fifteen-by-fifteen-foot room. But we failed. We began running in place, clapping and singing Airborne infantry songs. Florez sat on the bed, coached us on, and even smiled at us. We were making him happy, which seemed impossible at times. There was a lot of energy there. He said we were doing real good and that we should always sound like this, even outside.

Lights out in four minutes. Must stop.

7 January 1982
   Some notes on yellow pad:

- Classroom primary rifle training this morning.
- Temperature with windchill: twenty degrees Farenheit.
- Removed our BDU tops and did PT in our T-shirts.
- I made a point to get over on girls' side of Group B so that I could get some good views. There were some good views—with just a T-shirt on. Some of these girls have very nice bodies.
- Ran around the gym in three circles—this was pretty good. After many laps, the entire company started chanting, "One, two, three, fo-o-our-r" and clapping. It was very motivating and made some DIs smile.

   After PT, we got in our company formation again and started to march back. Sergeant Gibson started singing:

> "Way down in the valley
> I heard a great noise.
> It was mighty mighty Charlie
> Treating Alpha like boys.
> Whoa, whoa, whoa, whoa.
> Mark time, huh!
> Half left, huh!
> FLR, huh!"

   He was not satisfied with what we sounded like, so he dropped us right in the middle of the street in the FLR and made us pump out ten push-ups and told us we sounded like shit.
   We got up again and Gibson repeated the same songs, but he was once again mad and dissatisfied with our sound—we marched about fifty yards, and Gibson dropped us again.

Made us sing the song in the FLR and then made us do twenty push-ups. We sang better after that.

Sergeant Hensen has what he calls an "idiot squad." Those who constantly screw up their morning uniform, he makes them get their uniform of the day on and then stand in his office until we leave.

Things are getting edgy around here. We are starting to bicker more:

- Me and Pearson, weights—he rolled them at me.
- While standing in line, the one guy didn't want to wait after me while the door was closed.
- Last night, the fire guard tried to bitch at me for leaving a book, glasses, and magazine out. He kept asking questions, trying to find out something, but I just told him to mind his own business.

It's 8:00 P.M. and on personal time now until 8:45 P.M. Tonight we just reviewed the protective mask procedure—Orfino was the demonstrator. After Sergeant Hensen was through using Orfino as an example and said to him, "Okay, you can take off the mask," Orfino did and of course Hensen came on with his smug grin. All clear, all clearsky. Everyone laughed, but Hensen pointed out that the reason he does this is because the greatest disservice he could do us is to be real clear about everything and never try to trick us. He stated, "That would be real fun, wouldn't it? Everything would be real rosy. Well, let me tell you, the world doesn't work like that— the army is very unforgiving, very unforgiving! Life can be going real fine and lull you into a false sense of security, and then it comes up and hits you hard. Outside in the civilian world, somebody may screw up some statistics or something, and the worst that is going to happen is that they are going to get yelled at or maybe even lose their job. But that's about it.

The army is different from life. In the army, if one person is unconscious of what is going on, he could lose his life in the right situation, but if that isn't bad enough, one man screwing up may cause the loss of eight or ten, or even whole battalions. We lost whole battalions of men in Vietnam because of one person. One radioman can save or cause the loss of a large number of men. Discipline is needed. Knowing what to do is not going to help anybody if they don't have the discipline to do it. In fact, it is even better to have a disciplined person, because at least he will try to do it! If a man looks up in the air and sees a bursting starburst above and doesn't do anything but roll over and go to sleep, he could cause the loss of many others. If he was the only one to suffer, that wouldn't be as bad, but we are all dependent on each other. Our mission is to seek out, engage, and destroy the enemy. And the enemy's job is to seek out, engage, and destroy the enemy. And in that case, the enemy is *us*. They aren't going to be sitting on a powder keg waiting for us to come along and light it. They are serious, too."

Hensen said the Russians train with real poison gas. He said they lost 170 men in one battalion awhile back (last year) because their own army gassed them. Their battalion commander didn't even know! "Do you think you would be able to walk up on the survivors of that battalion?" he asked.

I was just talking to Snyder about Hensen before he made this little speech, and we discussed his somewhat sarcastic attitude at times, especially since he shaved off his mustache Monday. I said, "Yes, he is sarcastic much of the time, but I think he uses that technique on a professional level and not because of anything personal." I stated that I talked to Hensen on a more personal basis at times in his office and he relaxed a bit and talked on a very intelligent, congenial, philosophical, and psychological level. He is basically very intelligent and serious about his work.

Noon, just got back from lunch (hot dogs, cole slaw, orange juice, mashed potatoes). Marching drills before lunch.

This morning was used for review tests for protective mask, M-16 assembly and disassembly, load, clear, immediate action—reported to officer—and identify two ranks. Got weapons this morning at sunrise. Very cold and bitter.

Sergeant Florez acted as officer—before I got to his office, it sounded like he had a whip and hammer in there: crack, boom!

This afternoon we are going to confidence course. Weather is breaking—clear, warm, sunny, about thirty-five degrees out now.

At this point I am feeling rather insecure about being a responsible soldier. I know that at this time I would not be good enough to know what to do and how to do it.

Last night Hensen said the average size of recruits is getting smaller. I'm gonna talk to him about it.

It's 5:30 P.M. Ran confidence course today. Third Platoon of C-4-3 (my platoon) got the company trophy. We did very well.

The course is among the oak forest glens of Fort Leonard Wood. It was a beautiful cool day that we were out here at the confidence course—grand trails, and so on. The trail to C Course was obviously a marching trail, made of groves, had four trails on it from squads of a company. Along both sides of the trial were the DIs' trail.

We also went to Immunization today! Marched about one mile to another battalion gym where a nurse there poked our shoulders with a pin. We were all paranoid.

Song

Everywhere I go
There's a drill sergeant there.
Everywhere I go
There's . . . a drill sergeant
    there.
Drill sergeant,
Real sergeant,

Chorus

Why don't you leave me alone?
Why don't you let me go home?

DI

When I go to chow
There's a drill sergeant there.
When I go to bed
There's a drill sergeant there.

Chorus

When I dream at night
There's a drill sergeant there.
Drill sergeant.
Real sergeant.

Chorus

Company

When I dream at night
There's a drill sergeant there.
Drill sergeant,
Real sergeant.

Whoa—whoa—whoa—whoa.
Sitting home drinkin' Coke,
Thought my life was a joke.
Saw the army on TV,
Thought it was the life for me.
Whoa—whoa—whoa—whoa.

<table>
<tr><td>DI</td><td>Company</td></tr>
<tr><td>You get a line,
I'll get a pole,</td><td></td></tr>
<tr><td></td><td>Honey, honey.</td></tr>
<tr><td>You get a line,
I'll get a pole,</td><td></td></tr>
<tr><td></td><td>Babe, babe.</td></tr>
<tr><td>You get a line,
I'll get a pole.
We'll go down
To the fishing hole,</td><td></td></tr>
<tr><td></td><td>Honey, O baby mine.</td></tr>
<tr><td>Used to date a high school queen,</td><td></td></tr>
<tr><td></td><td>Honey, honey.</td></tr>
<tr><td>Used to date a high school queen,</td><td></td></tr>
<tr><td></td><td>Babe, babe.</td></tr>
<tr><td>Used to date a high school queen,
Now I date an M-16.</td><td></td></tr>
<tr><td></td><td>Honey, O baby mine.</td></tr>
<tr><td></td><td>Go to your left, your right, your left,
Go to your left, your right, your left,
Hey!
I want to go home!</td></tr>
</table>

You get an ax,
I'll get a saw,

You get an ax,
I'll get a saw,

You get an ax,
I'll get a saw,
And we'll cut the leg
Off my mother-in-law,

Honey, honey.

Babe, babe.

Honey, O baby mine.

Go to your left, your
    right, your left,
Go to your left, your
    right, your left,
Hey!
I want to go home!

Whoa—whoa—whoa,
To the hills to the front
We're going on a Russian hunt.

Whoa—whoa—whoa,
Russian soldier in my sights,
I squeeze the trigger; he'll die
    tonight.

Whoa—whoa—whoa,
See your friend lying dead.
He got a bullet in the head.

Whoa—whoa—whoa,
Who'll be the first on the block
To come on home in a box?

Whoa—whoa—whoa,
Fly all day,
Fly all night.
When we get there,
We're gonna fight.

Whoa—whoa—whoa.

9 January 1982

It's 5:30 A.M. Yesterday was supposed to be the end of total control—that's why we got to run the confidence course. It was fun, and we got to release a lot of pent-up energy that we felt because of the DIs. That's one reason we got to run it. As you know, we can't hit DIs or talk back to them. It also has a new meaning of teamwork and competition between teams. My Third Platoon gets to eat first for a few days. That may not sound like much, but when the alternative is icy winds or a crowded double chow line, it takes on new meaning. Lately our chow line has been double. I think the army is just givin' us shit.

Kawalski has been getting in trouble lately and adding a bit of humor for us:

1) FLR at classroom for sleeping for Florez—"I've been waitin' to git you."
2) For blocking traffic wrong, Sergeant Gibson. While we were marching, Kawalski laughed all the way and caused others to laugh that had to push-up to.
3) Not waking KP workers.

6:30 A.M. The weather is windy, cold, with snow, ten degrees Farenheit. Opposing forces classes today. Warsaw Pact, Russia (no Red Chinese forces?), all training aimed at Warsaw Pact.

10 January 1982

Fifty-five degrees Farenheit. Minus twenty with windchill. Fourteen degrees Farenheit is minus five with windchill. Long day today, very cold, clear, beautiful carmine sunrise and sunset today.

Went to church today, Polish priest. Church was packed with young trainees—I guess this mass is a welcome, needed change in this BT.

Phase 1 tests today on:

Put on protective mask	Go
Visual and vocal alarms	Go
Decontaminate clothing	Go
Decontaminate equipment	Go
First-aid for blood agent	Go
Stop bleeding	Go
Treat for shock	Go
M-16 disassembly/assembly	No Go
Load and fire M-16 rifle	Go
Reduce stoppage (immediate action)	No Go
Clear an M-16 rifle	Go
Rank ID/departure	Go
Report to an officer indoors	Go
React to an approaching officer/NCO	Go

What I have left is:
Attention
Parade rest
Right–Left face
About–Face

Present arms/order arms
Forward march
Rear march
Left flank/right flank
Half step, mark time
Halt

Lights out at 8:00 P.M. Tomorrow starts Range Week.

It was a long day. Mostly standing around a lot and waiting to be tested. The first test was to stop bleeding and treat for shock. That was great because it gave me a chance to sleep on the job as a victim. I almost got to be victim for the entire group but was not so lucky—Go.

The next was reporting to an officer indoors, to Sergeant Florez. It was his usual style. He sat behind the desk like a demagogue, slapping the rank covers and pounding the desk. I got a Go, but many others didn't.

After that was the M-16, first disassembly and assembly. Sergeants Cogan and Utley were in charge of this operation. And Sergeant Utley was ugly today. She gave me a No Go for not having my buffer and buffer spring snapped together. She got real cranky, yelling and screaming. Then she stomped my receiver pegs closed because they were turned upside down. No big deal, but it gave her an opportunity to get violent. She was very hostile today to everyone. She had to be on the rag today. I would bet at this minute that she was on the rag today—a real rag bag. Then she tested us on load, immediate action, clearing the weapon (M-16). She wasn't even watching me and gave me a No Go. I found out later that she did that to at least five or six others. We told Hensen about it but didn't get any sympathy. He just said, "Do it right next time." He knows this test at this time is not very important, and he is not going to risk damaging any authority of Utley just because she had a bad day. We think this test is important, though, because if we get three No Goes, we do not get our first pass. If we do good on this, it also will decide if we get off total control or not. Even after we were just about done, Utley kept yelling and screaming and finally put all of us (about fifty-five) in the FLR in the hallway. It was rather comical, because there was no room for us and most of us just knelt like dogs. It was like a giant game of Twister. Later, when we were upstairs, we heard Utley scream more and more and put them in the FLR.

After Utley, we were full of tenseness, and we had to greet an NCO and officer indoors. We were all tense and Sergeant Moran was in charge. He is a short black man and very relaxed. When I got in there, I performed perfectly, said, "Good afternoon, Sergeant," and, "Good afternoon, sir"—and even saluted perfectly. The only problem was I should have said, "Good afternoon, ma'am," to the lady officer. I just went through it the way I was always practicing but fucked it up. Sergeant Moran gave me a Go anyway. Bless his judgment.

After we were done, Sergeant Moran organized us and talked to us in a very encouraging way: He said we are just nervous when we say "ma'am" instead of "sir." He said we are just getting into the army and a new life and "Don't worry about these small mistakes. Don't let them get to you and make you lose your motivation and confidence." He said when you lose your confidence, you are absolutely lost—absolutely. "If you are getting headaches, it is just because you are thinking for once—there is a lot of fat around your brain, and it will take work to get it off. Relax a bit. You can do it"—and always in a patient, relaxed voice. He did relax us; an encouraging word in the times of stress is very helpful.

It's 7:50 P.M. This was our last night for cleaning the latrine. That is the worst cleaning job there is. About half of the people assigned, sixteen, do not help out. You can always count on a few lazy people in a group and a few energetic ones.

I had to clean washers and dryers tonight (assigned) but

also helped with sinks. Others' jobs were the washing of showers, sweeping floors, mopping floors, cleaning commodes, washing mirrors. Now we are assigned the simple job of the hallway—mop and buff. Easy.

Tomorrow we get up at 4:00 A.M. and I am not packed and ready yet. Tomorrow morning will be a rush.

Before going to bed, Sergeant Hensen called out, "Third Platoon!"

"Yes, Sergeant" reverberates.

"Tomorrow you get up at 4:00 A.M.—that's earlier than most of you get home at night. Good night."

"Good night."

January 1982

Cold, minus seven degrees Farenheit, minus thirty degrees Farenheight windchill—very hectic this morning. Finished our Phase 1 test:

- Got another No Go—in rear step.
- Called to go to range.
- Shipped there in crowded cattle truck.
- Shot at range.
- Sergeant Jones was our range educator—he gave demonstration, selected three individuals whose homes was in Florida, New Jersey, and Missouri (KC).
- Chose Karen Thomas. Demonstrated M-16 shooting and ammo box with water in it. Pretended it was the head of Thomas, told Thomas he wanted her to hold ammo box in front of his face—had crowd say, "Bang," and had Thomas duck.
- Had another sergeant shoot ammo box about ten times and blow it up. We all watched from a small grandstand that had about four pairs of bluebirds living in it.
- Other sergeant then shot rifle on automatic to show how

little it kicked. He then placed the butt against his knee, stomach, chin, and groin and shot on automatic to show how little the M-16 kicked. He had volunteers check him for padding, but they refused to check his groin. Then a volunteer shot the M-16 on his front leg. A very good demonstration.

Today Sergeant Jones said he is the only person in the world who can smoke on this range.

I shot today but failed to get my M-16 zeroed. Sergeant Reid and Captain Liter were really getting into Rentel's shit about the way he shot.

Today, chow brought in in a deuce and a half. We were the second platoon to eat, but I was taken out of the chow line for looking around. I had to stand and wait for the end of the line—also Sergeant Hensen assigned me to one to two o'clock fire guard. Third Platoon took first again in C Company because of Phase 1 test—average score 86 percent, next highest, girls, 82 percent. Sergeant Reid is happy about it.

Hensen said to me while I was in chow line, "Stanley, what are you doing standing out by yourself? Were you running your mouth?"

"No, Sergeant."

"Were you looking around?"

"Yes, Sergeant."

"Ahhhh—well, why don't you put your name in a prominent spot—about one to two in the morning fire guard?"

12 January 1982

It's 6:45 A.M. Went to range 2 again today in a cattle truck. It took me thirteen more shots to zero in M-16. I was getting very frustrated to the point of hopelessness—I finally got three shots in the white ring of the target. After I zeroed my

M-16 I went back to the classroom at the range and waited until lunch and also until I went on with the transition targets, which is eighteen rounds shot at six small targets—three at each. This is to check our zero and grouping. This was done in a heavy snowstorm. We got about three inches of snow today. And it is lightly snowing now. It made Fort Wood prettier and more like home.

On the cattle truck today, Sergeant Williams said the First Platoon (girls) is going to take the Phase Two test and if they don't he is going to kiss all of Third Platoon's asses.

Sergeant Jones at range 2 had some more tricks up his sleeve. He had said he had a surprise for anyone who leaned against his well or bleachers. Over the course of the day the sergeant caught three people doing that. His surprise was to give one and then two dice on up and make the convict roll them together and then do as many push-ups as showed up on the dice.

"Roll 'em," he would say in his deep voice and Mississippi accent.

"One, two, three, four, five, six, seven." He had twelve but would usually stop at six dice, pull out the other six, and then ask the sampler, "Do you think you can finish? You are half-done!"

They would usually do fifty or sixty push-ups. If a seven or eleven, then the contestant could quit rolling (as Otero did). "Roll yo own," Jones would say.

"One, six, ten, fifteen, twenty-one, twenty-four." It was very funny and surprising.

We did not get off total control today. Sergeant Gibson was pissed because people were not saying, "Sergeant," or, "At ease." So he issued a challenge—he gave us fifty points tonight to the company. One point will be removed by sergeants for any infraction—as of now. We are down to forty-seven.

13 January 1982

I am in full uniform now, waiting in my room to go to the rifle range. In the barrack these uniforms are very warm, confining, and bulky. Our shooting mittens are tied together by a long string that runs inside our coat sleeves. No inspections yet. Loud blacks in FLR by Florez.

Went to range sixteen—there about 8:00 A.M., waited around until 10:00 A.M. because we had no ammo. We waited in two snappy tents with a smoky stove in a can. Situation was like I expected it to be, like "MASH"—miserable tents, very smoky. I had to get out to get air. Very crowded inside, mostly blacks were huddled around the stoves. Sergeant Williams said blacks are 20 percent colder than whites.

The weather was snowing a bit when we got up—we got an inch or two last night for about a total of four inches that fell in the whole day (yesterday and today). The sky cleared up about ten and was sunshiney all day—The wind started blowing about 5 mph. It made it cooler. However, in the low winter run we managed to gain some extra warmth. We shot twenty rounds today, ten at 100 meters and ten at 200 meters. We shot at a life-size target of a man. The targets would be lowered to a bunker after each five rounds, and white markers were placed on the target to show how we were hitting. I had fairly good groups and was always in the black chest area.

Afterwards I had to go behind the bunker and mark the targets for others. It got real cold as the sun was going down. The NCO in charge down there didn't put on a field jacket or field pants or pile cap all day. I know he was cold because his ears and face were red. He just kept yelling and pacing all day.

It was cold. We marched back to the troop loading area and proceeded to empty our pockets for shakedown. It continued to get colder, and finally the cattle trucks came. I never thought I would be very happy to see one of those ugly crates.

And even happier for them to pack us in like cattle. No one was complaining tonight.

On our fifty-point challenge for our gain of total control we are now down to thirty-six points and it is only Wednesday night. We have until Saturday morning to go. We lost six points for a few recruits who were smoking in one of our living tents. And Sergeant Logan took four from us because when we were forming our platoon near the barrack many of us wouldn't shut up—and powee!

14 January 1982

It's 7:35 A.M. Lights out tonight at 8:30 P.M. We got up at 4:00 A.M. It has been a very long day, very long. After breakfast we marched three and a half miles in the dark and arrived at range 4. This was a pop-up range. The hike there was our first of this length, and it was tiring carrying our packs and rifles and in our heavy clothes. The skies were clear, and it was cold out, probably about ten degrees Farenheit or zero. The hike was along a troop track and through fields and forests. So actually it wasn't all that different from a stroll down some county road in Wisconsin. Tracks of deer and red fox showed up often in the fresh day-old snow. It was back to nature a bit, but not as relaxing as a solitary walk. It is rather difficult to be solitary among 237 other trainees.

We made it to the range just in time to see a beautiful carmine sunrise.

The range is a pop-up range consisting of three different ranges of man-sized targets—75m, 175m and 300m. We had 50 practice rounds and 40 testing rounds. So we each shot 90 rounds today—90 x 237 = 1,330, a lot of rounds for one day of shooting. It was fun, however. On my test, I got 34 of 40. Hawks was my partner, and he got 37 of 40.

The range instructor was a man named Sergeant Gonzales. He claimed he didn't speak Spanish (but he had a

Spanish accent). He said about twenty people from the other company asked him if he spoke Spanish. So today he just said no. He said he spoke Russian, French, Korean, and Chinese. He explained where men's and women's latrines were today, and he said if he caught any males near the females' latrines their asses would be grass and he would be the lawnmower.

Today was clear and cold, and early in the morning I had to run in place to keep warm. All morning I watched a warm front come in with its very high rippling and slicing clouds. Everything fast-moving clouds—very wispy and mosslike, very light, pretty. This afternoon the rest of the sky clouded up and the warm front took over. Temperature increased about twenty or thirty degrees—now at this moment, things are about twenty-eight degrees outside. Finally this weather has broken.

Served in mess line at lunch today twice—nuts, never volunteering again.

This Basic is getting long and tiring—we need some free time. Maybe we will get off total control this Saturday, I hope—I am getting fatigued and edgy; so is Snyder, so is Vieths. We are all looking to getting out of Basic and on to our AIT—Berella, Bang, Johnson, Vieths, me, and others. To California, what a great dream that will be. It just is that we never have time to relax except for sleep, and that is not enough. We have thirty-eight minutes tonight, but that is not relaxing. All we do is rush to write letters, shower, pack laundry, or something like that. We do not have the time to just plain old relax and dream away a waking day. I can think right now of so many people I know who are just relaxing and watching TV or something like that—making love, snacking. We also talk and think much about women here.

Sergeant Gibson says he is getting edgy—he always does around range quality time. He says all the senior DIs come and watch Charlie 'cause we're always the best. The senior DIs compete for the best company—among each other. He says they are not supposed to, but they do. I'm willing to bet that Gibson and the others have a big wager on it. Gibson says that how a company does reflects on how good the DIs are.

By the way, on the first Phase 1 Sergeant Byrd was checking score sheets and saw everyone had Goes. He said, "Y'all are making us look good."

Johnston forgot his overshoes today after Sergeant Reid told everyone about three or four times. All the DIs were very pissed. They said they were going to give an "A" 15—because if he gets frostbite, all of the commanders blame the DIs, not the private. "Cover your ass" is the way of life here.

Also, Jefferson got a $128 fine and fourteen days of detail—"A" 15—for disobeying an order of Sergeant Reid's.

Menu tonight: roast beef, sauerkraut, milk, lettuce, bread.

For lunch, spaghetti or pork chops and potatoes plus bread, Hostess pie.

Tomorrow we get up at 4:00 A.M. again and go to range 20—pop-up targets in shooting cages (forest).

15 January 1982

Get weapons, weapons court. Road march to Combat Range 20.

- Find targets in woods.
- Estimate distance.
- Shoot combat range.
- Shot 32/40, pop-up targets.
- In a shooting line in the woods.
- Left mittens at foxhole.
- Shoot from prone and foxhole.
- Very warm in classroom, almost fell asleep.
- Mail from Stuart, Mom, credit union.

Weather got up to thirty-five degrees Farenheit, very warm today.

16 January 1982

Got cold out again, minus thirty-five degrees windchill—a new front moved in.

At 7:00 A.M. it's about fifteen degrees out now, windy. No doubt it will get colder.

Baez told me she is pregnant today at breakfast (not from me). She is getting a discharge—her teeth are falling out too. She is not going to keep the baby because she has had seventeen X rays and gas chamber and shots and drugs—says there is too much chance of deformities.

Return to classroom, clean weapons.

It's 6:25 P.M.—notes about rest of day. Meeting in classroom (Bat HQ). Captain Liter took us off total control as a challenge to us—everyone cheered. Liter placed a yellow banner on our company flagmast, said the weather here is the coldest in 100 years.

We can expect at least one more week of this. Liter is happy with the way we sounded off with our company motto.

> "The difficult we do immediately,
> The impossible takes a little longer.
> Miracles by appointment only.
> Duty, honor, country,
> Drive on, drive on."

Captain Liter was rather perturbed about the fact that two people forgot to bring their overshoes to the range—he told how we could get frostbite from that. He also said if any of us forget any of our gloves, scarf, weapons cards, overshoes, pile cap—they will march back to their barrack from the range (four or five miles) with their buddy (who sleeps above or below you in your room). He said we are not taking care of each other enough and that this will ensure that you and your buddy will take care of each other. (Dan Vieths is my buddy.)

One out of ten officers is a West Point grad. Eight out of ten generals are West Point grads.

Sergeant Gibson talked after Captain Liter—he also talked about total control, our Phase 1, and our work here, self-discipline.

Gibson said in his own proud, gruff sort of way as he strutted across the stage, "You sounded real good last night, Charlie. That's why I marched you around Two and Three Brigades last night—that's enemy territory. And I want them to see us. We are the best company on post—I ain't bo-shittin'. Everyone knows it—it's the truth. That's also the reason why I took you by your sister company, B-3-3; you know I don't like them. We are the best."

Sergeant Hensen said that if he had a sister in a brothel or a brother in B-3-3, he would rescue the brother before his sister.

These notes are really choppy here and also the last few days. I am having trouble concentrating and getting in a creative mood. I think it is mostly because of the fact that I never have any time to just *relax* in the afternoon or evening in my bed—not sleep but just relax—and think and reflect. I never have any time to just relax. I suppose now that we are off total control, I will get a little bit more time to relax. Sergeant Hensen took us over to the brigade PX—some of us had some beers (not me). It was smoky and many of us were just talking and eating candy. It was only for an hour and a half but a good break. We didn't get any passes today or able to use the phone because they do not want any cases of frostbite—they (sergeants) say if a private gets frostbite, the private doesn't get bitched at, but the DIs do! So to prevent this, they do not let us outside for long. I am used to much colder weather than this, and I know how to handle myself in it. But obviously they are worried about all of the people from the South who do not know how or when to get out of the cold.

Today instead of passes we practiced reporting to officer indoors, reporting to inspecting officer on guard duty, inspection arms of M-16, prone unsupported shooting position, and LAW (Light Antitank Weapon, a small bazookalike weapon that is used once and discarded). Sergeant Reid says after you shoot at a Russian M-60 tank, you had better put that turret out of commission or it will frazzle your ass with its gun that shoots a mile a second—it will take about four or five rounds of LAW to put out the tank. Russians duplicate LAW exactly for their troops as usual.

18 January 1982

It's 7:30 P.M.—lights out in thirty minutes. Today we went to range 21, the last combat range with pop-up targets before Record Fire tomorrow. I shot 36 for 40, good enough for an Expert rating, but I will have to do that tomorrow. We shot 50 rounds, 10 were practice, 20 were shot from foxhole, and 20 from unsupported prone position. It takes only a few minutes to shoot all of the rounds but we build up tension all day for it . . . and then it's over.

Today the weather warmed up even more, to around forty degrees, so that made shooting better, too. The sun shone all day brightly and helped melt away the remaining snow. The area around the range is typical Missouri oak forest and probably very pretty in the summer and fall.

Our range instructor today was Sergeant Shipley. A Vietnam vet, thirty years old but looks younger, very good shape, was Special Forces in Vietnam, also a DI in South Carolina, is proud of his bachelorhood—talks about beer and pot, says he was shot three times in Vietnam while he was drinking beer or smoking pot. Says lead is hot and burns—was shot by snipers each time.

Shipley tells Polish and Bohemian jokes: How to separate a Polish man from a Polish boy—use a crowbar. What is the most dangerous job in Poland? Riding shotgun on the garbage wagon. Both are in poor taste. He also asks who is from the import capital of the world—Miami. (It is sick and stupid to romanticize drug smuggling.). He also asks who snorts cocaine.

Tonight Sergeant Gibson gave us a reprimand speech—mostly to smokers who are smoking without permission.

- Says he will put us back on total control.
- Says our mess hall manners will remain the same.
- Says we are embarrassing him and C-4-3 by not saluting the captain.
- Says if we do not start doing that he will suspend passes to keep us from running into brass at the PX.
- Says he does not want to see "C-4-3" on shithouse walls. He don't give a damn if he sees "B-1-1," "C-3-3," "D-1-1." "We do not do that shit here, and we are not going to allow you to do it."
- Says someone from Fourth Platoon brought beer to the barrack from the PX—says that will be $100 fine, says if he catches anyone smoking in the barrack that will be a $50 fine.
- Says most of C-4-3 is doing well, but there are a few in each platoon who are going to screw it up.

Sergeant Hensen said we are doing well and we should keep it up—said we should relax tomorrow when we shoot for the record and remember what was taught us.

The brass (officers) said: And we will do good—get comfy.

This morning I saw recruits who were shipping their BT gear back to the warehouse at pickup point. They were very happy and wild—they only have three days to get out of Basic. It made me happy to see them and something to look forward to in one month. Their DI walked over to Sergeant Gibson and said, "These are the worst I've had." Gibson said, "Hell, you never taught them any discipline. It's your fault." McCubbin and Pearson really started pushing each other—they constantly antagonize each other. We had to break them up.

20 January 1982

I am writing at 7:20 A.M. I didn't write about yesterday at all, the nineteenth. Yesterday was a very long day. Got up at 4:00 A.M. and left around 5:30–6:00 A.M.. We marched to range 1 (Record Fire), 3.5 miles, and shot there until noon (34/40). Then we marched a half-mile to range 3 and shot Automatic Fire until five. Then we marched back to range 1 at chow (dinner). After dark we shot night fire. We ended up getting back to the barrack at 9:00 P.M.—very long.

Last night we got nine hours of sleep. They must have tried to be nice—my, yes. We didn't get up until 6:30 A.M., whereas for the last week we have been getting up at 4:00 A.M.

At Auto Range we shot three rounds on auto and then an eighteen-round clip in three-shot bursts. This was a real experience here.

The sergeant told us not to do what another guy did once. "When everyone started shooting on automatic, he picked up a couple of handfuls of gravel and threw them at the outdoor classroom" (with three aluminum sides and a roof, open in the front, housing the bleachers). "It sounded like machine gun bullets hitting the building, and everyone scattered in panic."

When all of us in the firing line opened up it naturally sounded like a war, but for anyone who hasn't been to a war it also sounded like an army of streetworkers pulverizing the concrete with jackhammers.

Night Range—we first ate lunch outside again at dusk. We were all tired bundles of green quietly eating at the green picnic tables in the fading light. Looking at the sight, someone would suppose we were relatively harmless, with little on our minds but our stomachs. And they are right, too—here. It is an irony that we as fighting men are just quietly minding our own business like a few rabbits on the wet morning lawn.

21 January 1982

Weather at 6:00 A.M. pouring rain. Last two days have been cloudy, misty, cool (about thirty-five degrees). Forecast today is for freezing rain.

Yesterday, Moore and McCubbin got into a pushing fight—both were hauled down to Sgts. Reid and Hensen's office. There have been small skirmishes in our platoon, mostly between whites and blacks. Not trying to be biased, but it seems mostly like the blacks' fault. They are overly loud, swear constantly, have no or very little respect for others' rights, and if anyone tries to correct them they get all defensive and take it personally. As a result of Moore and McCubbin's bickering, they have to be buddies now and do everything together—fire guard, hikes, etc. Most of us think it is humorous—Hensen is shrewd.

Other small incidents that piss me off:

Ask blacks to whisper in their conversations, and they find that impossible and continue to yell boisterously.

While mopping the floor, I asked a few persons to remove their muddy boots. I kept asking but it didn't help. I finally gave up.

Last night Sergeant Reid gathered us into the classroom and told us as a platoon we have a seriously big problem: "I'm talking about racial prejudice overtones. And it is so thick here we can cut it with a knife. I don't like whites, blacks, Russians, Jews, Japs, or anything else, but I love people; this thing cannot keep up. If it does, you're going to go down the road. A house divided will soon fall. I think that this country has the biggest and best-equipped army in the world, but it is way behind in discipline and is too easy. This is my last cycle, folks, and I'm going to train disciplined soldiers, and you'd better realize that now. If you do not like blacks, whites, Puerto Ricans, you are going to be in for a lot of heartache, disappointment, and problems, because that is all there is in the army. If you do not like these people, then you get out now. Get out now, please. It will save a lot of trouble in the future."

Vieths feels tensions are caused by the large intake of negative input and no place to let it out. Due to the weather, we have had very little PT. That is another cause by itself—the weather. Now it is foggy out.

Yesterday (last night), Sergeant Reid put Hawks and Ogle into the FLR for fifteen minutes and gave them three nights of fire guard for being three minutes late for a meeting he called. Hawks and Ogle were laughing in the FLR.

Yesterday, 20 January 1982, we went to orienteering course—the Valley of Big Piney River, bluffs, ridges, and grassy valley. Looks like a good place to raise horses. It was beautiful out here. A quarter-mile-wide valley will have about 200-foot bluffs. It reminded me of the La Crosse area in ways, except the valley wasn't as wide or bluffs as large. The timber was oak.

We had a little class about the orienteering course on a bluff overlooking the valley area. The course was in the valley (grassy) and consisted of seven points that we had to find. The points had map coordinates on them and gave coordinates for the next point. We would then check the coordinates on the map, plot our course, and then find the next point. All were near the road and so it was no problem and all within about a half-mile of each other. It was very easy. I was disappointed that the course was not out in the wilderness like I thought it would be. I wanted a chance to prove my prowess and skill in orienteering and landscape reading but was cheated out of it. It was nice, however, to get out in the country and get some fresh air without the DIs constantly over us. They were cruising the roads keeping an eye on us so we wouldn't get into trouble, but it was not like they were right over our shoulders. We could be groups of four trainees just walking across the fields looking for points and generally having a good time. One point was located on the banks of the Big Piney River (which is only about twenty-five feet wide). The water was clear and bubbling and was a pleasure to see.

DI yelling at a trainee. Trainee pulled out wallet and opened it and said, "Beam me up, Scotty; it's getting hot down here."

I finally got both pairs of boots cleaned up and polished today—they were all muddy and hard from two days of marching in the wet mud. I have been learning how to be much more conscientious about what I am wearing and the condition it is in. I feel uncomfortable with something unclean.

23 January 1982, Saturday

This morning had class on military conduct of POWs—name, rank, and serial number; try to escape immediately after you are captured, because your chances of a successful escape are greater in the area of capture, as you are closer to friendly troops. Once you are behind enemy lines, your chances of escape are greatly diminished. Much of this is repeat from lecture about the Geneva Convention. Another segment of the class on conduct by Sergeant Florez was the question: "Are you prepared to give your life for your country?" He asked the question of our company and a good part raised their hand—and he said in a doubtful way, "Are you *really?*" He said many are in the army for just the money. Many are pencil-pushers and are not prepared to die for freedom. He said also many of us will be going to good training schools, but if an emergency broke out, Uncle Sam would pull those books out of our hands so fast and replace them with ruck-sacks and M-16s so fast that your head would swim.

I personally did not raise my hand when he asked, "Are you prepared to die for freedom?" Nor did I raise it when he asked, "How many are not prepared to die?" I would go willingly to any front that the army ordered me to and also do my best to fight, but I don't know. I don't think I am *prepared* to die. I would die, true, if fatally wounded, but whether or not I was *prepared* to die is another question.

Florez showed a slide of Nathan Hale on the screen with his famous statement: "I regret I only have one life to give for my country." His age: twenty-one. So young to have made a statement like that. How many today could honestly say something like that from the heart, I wonder.

Later morning, our second PT test—I did fifty-six sit-ups, up from thirty-three the first time, and forty-seven push-ups, up from forty-two the first time. Ran two miles in 14:15—last time did 6:44 for one mile.

I basically was satisfied by my performance today—basically, but I know there is room for improvement. I would have been much better if we could have done more PT since the first test, but we had a two-week Christmas break in which I did very little, and also since coming back to training nearly three weeks ago we have run only about four miles. The weather has been so cold. Speaking of this crazy weather, it's at it again. Today about 20-mph winds and temperature about fifteen or twenty degrees Farenheit—yesterday it was cloudy, misty, and warm. This is the strangest weather I have ever seen.

Today at PT we were standing in our lines waiting to do sit-ups (about twelve in a line) and Gibson sneaked up behind little Rentel and yelled real snappily, hoping to scare Rentel, "Why are you looking around?" Rentel nearly jumped out of his boots and stiffened up and didn't move a muscle but had a scared look on his face. Gibson just walked away laughing and digging his heels and walking like a cowboy. Everyone chuckled at this and then Sergeant Gibson came back to Rentel, put his arm around him in good-old-boy fashion and said, "Aw, nobody's gonna hurt my R2D2."

Moore and McCubbin have a three-hour pass today. But they have to spend it together.

We were assembled in classroom after noon chow, and Sergeant Hensen said we have six hours free today. All our mouths came agape at this miracle—he just warned us about coming back drunk (PT!), told us to watch ourselves and do not get him in trouble. Six hours—amazing! I will finish this writing, finish my laundry, and then go on my hunt for Patty Grippen with my spit-shined boots and my battle dress uniform.

Joe Hawks and Mike Snyder only have two hours for pass, because they were talking in chow across aisles.

24 January 1982, Sunday

Waiting for chow, 6:45 A.M. Just a short bit about yesterday on pass time. I went to the Little PX and talked with Alan Berella, Debra Colter, and another girl, named Harrison. I had a very good time. I found out that Colter is an amazing person. She is twenty-nine and will be going to Monterey along with several of us. She lived in Ireland for six years and has written two books. One is a series of essays about Ireland, and another is a collection of poems she wrote. She made about $15,000 or $20,000 on the essays, I believe.

We stayed at the Little PX for a while—until about 6:00—

and then we took a taxi to a pizza restaurant next to the David Club. Harrison bought us a pizza and beer. We sat talking about drill sergeants and who they reminded us of in stardom.

Redhead Smith tried to kill herself tonight.

Sergeant Gibson bitched at us today, said we did bad yesterday, said there are about fifteen of us that fucked up, said guys and girls were in the stairway talking to each other in their panties and underwear. "This ain't no whorehouse. I don't care what the hell everyone does at other companies. We don't we do that shit—we don't do that shit. I'm going to get rid of you people, now, this minute. We don't want you." (He named about ten names and told them to go to the back room where he could yell at them.) "Hell, I just got done telling you all how smart y'all are. This is the most educated bunch of people Charlie Company has ever had by a long shot. And then you pull this shit. And you know what else? Today at breakfast, something happened that I haven't heard about since grammar school. Someone in this company was throwing grapes at the women in the chow line—throwing grapes. Hell, we have eleventh grade high school students come here in summer to do their BT and then their AIR after graduation and they are more mature than you. And you are grown men and women."

Don't nobody even talk to me, nobody!

Don't nobody say a word to me!

Sergeant Hensen gave a very humorous and informative class on personal hygiene today—very good!

25 January 1982

Sitting in mess hall with Caddel and Osterholm eating breakfast, and Carole King came on over the intercom singing "It's Too Late." It must be a mistake or else we are all dead and in heaven.

Commo Classes Today

- About field telephones—one is voice-powered and will go four miles with mikes.
- One is battery-powered and can call everywhere. Sp5. Martin called from the field in Germany home to his mother a couple of times. His mother did not believe him, and he had to have his platoon sergeant come out and talk to his mother and verify. Call was via satellite. He suggests getting to know the signal people in your unit in Germany well so they let you call home free.
- Another class with Sergeant Lloyd, this one about the field radio. Its range is five miles, and it has over 1000 channels— we picked up the Donahue show while we were learning how to operate it.

In a mixed class of men and women, Lloyd taught us how to work it and set it up properly:

1) First put magnesium battery in. This is what you will be tested for at 204; if you don't you will get a No Go and will be teed off and it will be *your* fault. (Finally something is mine around here.)
2) *Second* you put in antenna, the flexible antenna. If you use the ten-foot folding antenna, you will be *wrong* and you will get a No Go and if you get a No Go you will be teed off and it will be your fault.
3) You adjust the frequency; first you adjust the band—high or low—and then the first two digits of frequency and then the last two digits with the other knob. If you do not do this, you will be wrong. You will get a No Go. You will be teed off.
4) After that you hook up the telephone, adjust the volume, turn it on, and you're ready. If you don't do this, you will

be wrong, you will get a No Go and you will be teed off!
And it will be your fault.

"*Now,* class, what do you do *first* when getting ready to transmit?"

Tumultuous answer: "You put the battery in," Sergeant Lloyd.

"What if you don't?"

"We'll be wrong!"

"And what will you get?"

"And we will get a No Go, Sergeant Lloyd."

"And what will you be?"

"And then we will be teed off, Sergeant."

"And whose fault will it be?"

"*Ours,* Sergeant Lloyd!"

Finally we have something that is ours, something that is mine—in this army of sergeants who own everything. Everything is the sergeants'. They also say:

"Don't touch my wall!"

"Don't touch my railing!"

"Get off my grass!"

"Get out of my barrack!"

"This is mine and that's mine."

"And *you* are *mine.*"

Well, we finally have something that is ours, something we can control, something we cannot have taken from us—*our FAULT.*

We got back into Sergeant Lloyd's classroom that afternoon—a classroom converted from a 1941 barrack that was built by German POWs. I got Goes on phonetic alphabet, numbers 1 to 0, hook up radio, and hook up powered telephone. He looked around the classroom this time—with its stormcloud gray picnic benches for tables—noticed there were no women, then asked if there were any females. No answer at all.

Then, "No females, Sergeant Lloyd!" (tumultuous answer).

"Good!" yelled Sergeant Lloyd. "From now on, when you are wrong [arching his back and arms in the air], you will still get a No Go, but now you will not be teed off—you will be *pissed off*."

Tumultuous applause.

Today at PT we formed new tracks—A, B, C. I was put into Track A as a result of my performance in the second PT test—I did forty-seven push-ups, fifty-six sit-ups, ran two miles in 14:15, not real fast but better than most. It really felt good to get into Track A—it is not a real big accomplishment, but I at least succeeded in something and to some degree earned it. I could hold my head up higher and call the cadence of Track A with pride. One cannot help feeling a little bit better to be recognized and gain some much-needed self-respect. I did not make superjock, however, but now have a much greater desire to earn it and also max the next PT test, which needs 300 points—that is achieved when one does sixty-nine push-ups, sixty-eight sit-ups, and runs a 13:30 two-mile. I do desire it. I wish we could have gotten much more PT than we have gotten since Christmas vacation—it has been so cold, too cold to exercise.

Saturday, Angie Matthews came onto our floor looking for a buffer—I was in my underwear and she saw me.

Sergeant Logan, female, walks all around our barrack while we are nude or in underwear. She doesn't care, so I guess we shouldn't care.

Tonight I finally got the time—personal time—when I could use the phone to call home for stamps. I have been writing for two weeks to receive stamps from home—but *no*. I am desperate. I wanted to call my sister Lindsey—I asked Sergeant Reid if I could use the phone, and he said, "Ask the CQ [Sergeant Logan]." I asked Logan if I could, and she went into a pseudo-rage and said I was abusing my privileges, asking if I could call home for stamps. I thought the reason to call was irrelevant—denial was ludicrous. I turned away, laughing at the reason. It seemed too strange.

She then yelled, "Get back here! Do you have an uglier-ass smile on your ugly face than you had already?"

"No, Sergeant Logan."

"Are you laughing at me?"

"No, Sergeant Logan." (I was laughing; it was so ludicrous.).

She then went into a story on how she would take the whole company out and work us at PT.

26 January 1982

Confidence course—to be filled out on pass time.

Sergeant Gibson made Hawks run around company formation while we were marching, made Barnes, Lamar, Marshall run around, too.

Today we worked on review of Phase 2 testing. During this time four people from each platoon were chosen to go to the grenade range with another company to qualify. Gibson came upstairs to Third Platoon and said, "I need four smart people." Hensen immediately said, "Kawalski, Stanley." I don't know if Gibson was using "smart" facetiously or not, but we went downstairs. Said they wanted us to be makeups for grenade range, that next Wednesday while we are on bivouac we will need KP people, bivouac guards and so forth. I think that there is another ulterior motive of DIs for wanting us to be qualified. I think it is to get them out of work next week, like they tried to do with us with rifles (and then tell us we can get a medal).

Squad leaders switched tonight. They lasted five weeks, the longest Hensen has ever seen—Barnes, Johnson, and Chasen. Barnes says he is not happy about it, but everyone knows he is eating it up. Callaway, Hansen, Sims, and Kelvey are replaced, and they are all happy, especially Callaway. People in the platoon have not been doing the work he *asks* them to.

27 January 1982

This morning at 5:30 Barnes said while he was at the sink shaving, "Yeah, before I joined the army, I wanted to organize a safari to Afghanistan and get the head of a Russian colonel and put his head on my wall. Heck, everyone else goes trophy hunting. Persian ibexes, African gazelles and gnus, mountain sheep. Why not?"

28 January 1982

A very long day—lights out will be in twenty-five minutes.

We got up at 5:00 A.M., ate breakfast. Thought it might be a bad day. I heard Sergeant Hensen say that he only got one and a half hour's sleep last night. Found out I was right—Hensen was in a bad mood. We didn't get the word to get ready for formation this morning and about twenty of us (at least) were marked down in his little book and he said we would be doing detail this weekend and lose our passes and also be doing fire guard.

At this very moment, Sergeant Florez has his entire platoon in an FLR—they were getting him angry earlier; they wouldn't come and get their mail. Florez still has men in FLR—many are lying on the ground and are exhausted, are arching their backs trying to stay up. I just keep hearing them say, "Yes, Sergeant Florez."

When we marched over to get our weapons, I didn't sound off to the cadence, and many others didn't either, because we were so angry. We all felt Hensen's decision was unfair, because he made a poor announcement. He has been on the rag lately, and all of us agree that his reactions only make things worse, because they increase the hostility of us men. We marched to the loading area and I noticed it was warm out and a clear dawn sky. I commented that to Vieths, and Sergeant Gibson heard me say, "What a nice day." He evil-eyed me and yelled, "Did someone put you at ease? Did they?" "No, Sergeant Gibson." "Then git down and knock out ten."

We were loaded up in the cattle trucks out to range 30 Alpha—we were to receive training on offensive day assault with weapons and grenades. We received a small class on instruction.

- Ready.
- Moving (roll) run.
- Step prone position.
- Ready.
- Look—*movin'*.

This goes on until the enemy target is sighted, and then the team leader says, "Throw your grenade."

Partner: "Preparing my grenade!

"Throwing my grenade."

Team leader and partner hit prone position when grenade is thrown, "Grenade!"

Team leader repeats, "Grenade!"

We then got Third Platoon together to practice with a big, tall black sergeant, Sergeant Jackson, who was very laid back and eased up the tense atmosphere that Hensen had created. The Third Platoon took first in Phase 2 test by a long shot. Sergeant Williams's prediction was that the First Platoon would win and if the Third Platoon won he would kiss all of our asses. He didn't, though, but we rubbed it in a bit.

Sergeant Bolen says the name of the week is Goldsmith. (The name of the week is someone who is screwing up, and the DIs watch him constantly and harass him, especially in the lunchroom.)

We became mud pigs today when we low-crawled, back-crawled and rapid-crawled under three sets of barbed wire twice (once with gas) through three-inch-deep mud—nice warm day, but everything melted.

We were pigs—red mud—were also treated to more dirt while we threw eighteen grenades at three ranges at bunkers.

Tonight at chow, four of the white men were sitting at the table, four blacks at another. A Puerto Rican sergeant, Sergeant Marianda, made one white and one black switch places, saying, "Too many whites sitting together and too many blacks sitting together."

I said to McCelvy, "I didn't know you were black."

Osterholm said, "You do *now!*"

29 January 1982

Got up at 4:00 A.M. Got paid today, $436 out of $642 gross.

Many got anywhere from zero to $1,000. Sometimes computer screws up and overpays someone by a couple of months, and then after a couple months you will not get paid—you can see it makes for great planning. We stand outside the gym in a formation and wait for our call; then we report and say, "Sir, Private —— reports for pay," and show our ID. I figured that out of one hundred or so troops about $500,000 was put out that morning.

Later in the morning, we were picked up in cattle trucks, alias pig trucks. We marched over in yesterday's dirty field clothes caked with hard red mud—a dirty cloud hung over the moving platoon like smog and reminded me of Charles Schultz's *Peanuts* character, Pigpen.

Today started out quite different weatherwise—cold and windy and gray, much different from the clear blue skies and sixty-degree temperature of yesterday. Sergeant Gibson came out and said, "See, folks, you just can't plan on anything in this Missouri weather." We were picked up in the cattle trucks and taken to range 31 Bravo. (We went to 31 Alpha yesterday.)

We threw one live grenade each and that was exciting. We also qualified for grenade course. I got a sharpshooter score. The course was a veritable quagmire of mud—we had to throw grenades at six targets while wallowing in mud. My M-16 got all jammed up from mud, and also my magazine got stuck in the rifle. It would not come out, and I had to beat it out with a rock.

Hensen called out later, "Who had that magazine stuck?" (I knew he knew it was me and was just trying to make me feel small.)

"Me, Sergeant Hensen." I raised my hand.

Hensen asked, "Did you get that magazine out?"

"Yes."

"How'd you do it?"

I answered, "With a rock."

In the morning we were told about the different types of grenades: 1) incendiary and 2) smoke—green, red, white, and yellow. A Hispanic sergeant told us a story about the incendiary grenade—he says when we want to get revenge with Sergeant Florez we can take two incendiary grenades and put one on his front seat and one on his car's engine and that will incinerate his car. "I know you all do not like Sergeant Florez because he and I are good friends and he tells me all the shit he does to you." (He laughs.) "How many do not like Sergeant Florez?" Most of us raised our hands.

Sergeant Aberdeariz then started telling us another story.

He said he used to be a marine, a jarhead. "That is how I got started in this business. Down in El Paso, Texas, me and a buddy of mine talked it over for a couple of weeks and we decided to join the marines. We figured we would just sit on our asses and earn some money; what could be better? At that time I had a very beautiful girlfriend; we were going to get married. I told her that I would join the marines, go to Vietnam, and when I came back we would get married. So my buddy and I finally reached the end of our delayed entry program, so we went to El Paso together to sign the dotted line and swear in. The marines took me in. The little room first and I swore in—then it was my buddy's turn. Right then he chickened out and didn't go through with it. I went to Vietnam and he went back home. About one year later I received a letter from my girlfriend saying she was going to marry my buddy. I wasn't that teed off about that, but he took my car along with my girl. So, when I got out of the marines I looked them up and did to my car what I told you to do to Sergeant Florez's car earlier."

The incendiary grenade reaches a temperature of 10,000 degrees Farenheit very fast and burns for a short while. Sergeant Hemeraziz demonstrated one for us and very rapidly burned up a metal ammo box that the grenade was sitting on. I could see where it would burn up a car rather easily.

30 January 1982

It's 4:00 A.M. Supposed to rain all day today, all day—our fifteen-mile road march was postponed, and we are all thankful for that. Instead we are getting ready for bivouac and also have five hours of free time—also got some equipment for bivouac. Our trenching tool (fold-up shovel) was described by Sergeant Hensen as an all-purpose tool, a shovel, an ax, a scraper, "but it cannot be used for a weapon, Barnes."

Class laughs along with Barnes.

At 7:00 P.M. weather has changed from rain to heavy snow. We are back from our pass—about four or five are very inebriated, and many others are feeling good. Sergeant Hensen is displeased, very displeased, with our performance tonight. He says we are walking around like animals. Some are in the showers. When he wants a formation, he wants it now, not in five minutes, not in ten; the formation is not in the shower, not in the rooms. The main drunkards at this moment are Rentel, Downs, Keller, and Barr.

At 7:30 P.M.—Rentel, alias R2D2, was being led by Callaway and Paxton by the shoulders down the hall. R2D2 was staggering and bleary-eyed. They were trying to get him to his bunk; they had his head under the shower to try to sober him up. Keller was heaving in the toilet while it was flushing and vomited out his two false front teeth. He now has a distinct lisp. Barr, the guy that has been having trouble with his wife, two days ago told me his wife pawned his and her wedding rings for some fast cash and also told him that their love isn't very strong. I didn't tell him, but I think she is just playing games with him, trying to reinvigorate the relationship.

31 January 1982

It's 0200 hours—heavy snow outside. I'm sitting at the platoon table in the hallway in front of Sergeants Hensen and Reid's office—I am on fire guard duty again. I am on duty with Hopkins and with Mike Snyder. This is the first time that there have been three of us on duty at one time. The duty is not difficult. Here are our fire guard orders:

1) "Your post is a roving post."
2) "Your duties as the fire guard are to act as both fire guard and security guard. Your tour of duty will begin at lights

out and will last for a period of one hour. Fire guard ends at first call."

3) "During your tour of duty, you will conduct as a minimum four fire and security checks. During these checks you will physically check every lock in your platoon area. If a wall locker is found to be unsecured (open lock), the individual will be awakened and told to secure his/her locker, while you watch them. You will also ensure that there are no keys hanging on the end of bunks, under springs, or in boots, that there is no personal property left out except footgear. (Tonight, however, there is a lot of gear out because we washed much of our mud-caked clothing yesterday.) That there are no fire hazards such as smoking in bed or in the barrack."

4) "Should a fire be discovered, you will alert the barrack by moving to the nearest local fire alarm located in the barrack hall and activate the alarm, at the same time giving a verbal alarm. Ensure that everyone is awake and knows in what area of the building the fire is located."

5) "You will not quit your post until properly relieved by a replacement or your platoon sergeant."

6) "These orders are issued by approval of the unit commander. Failure to properly carry them out could result in disciplinary action under UCMJ."

Those are our directions. Not difficult.

Last night after Sergeant Hensen gave us a lecture about many of us acting like animals, he told us that the people of the night before last would be his fire guards again. Those were the people who were late for his formation earlier in the week, the morning after Sergeant Hensen had an hour of sleep. Well, he told us to put our names up on the fire guard roster, an erasable chart with columns for names of fire guard, the time of his duty and spaces to check in and out. When

Hensen went back in his office about sixteen trainees pounced on the chart and began clawing their way at it, trying to get to the "choice" time slots. The tearing and yelling to get a good spot was unbelievable—it was like watching a bunch of kids, all out for themselves. This to me is a good example of what happens when people are only concerned for themselves. I went to lie down for five minutes so the tangle would clear. After relaxing a bit, I went back out and was shocked to see that it was still going on. Barnes was patiently trying to get some order to the pandemonium—"At ease, goddammit, at ease." Meanwhile Hensen was beginning to simmer in his glaring way, in which he silently tells you something had better be done fast or the shit will hit the fan. Calmly: "This is the last time I'm telling you people to 'at ease,' or else.

Barnes, with frantic eyes, yelled, "Goddammit, at ease!" Finally they backed away and the spoils were divided.

It's 8:00 P.M. Twelve inches of snow last night. Base is temporarily closed. New, fresh snow makes me feel more at home, makes me feel like this is winter in Wisconsin. Conifers are dropping from the weight of the snow. Everything is pure white.

People are in better moods also because the outside looks so pure. The church bells chiming in the morning also make people happy. I got a 4.5-hour pass today. Great! Time to prepare for bivouac. We are glad it snowed; camping will be much cleaner and beautiful. We heard a rumor tonight that bivouac may be canceled. I hope not, because I feel bivouac is a big part of army Basic. I would feel cheated if I had this miserable experience taken from me. I value my miserable experiences very much because they make me laugh in the future.

I am getting fed up with taking shit from other privates, most notably blacks. I was in the PX tonight, and many blacks—big blacks—were butting in line to buy beer. They wouldn't listen at all to me when I pointed out their selfishness. They don't care at all. Why should I care about them black asses? Should I ignore this? Should I ignore my rights? Should I "turn the other cheek"?

1 February 1982

Monday morning, 0700 hours, waiting for our orders for bivouac. We are on hold until 0900 hours. The sun is not up yet, but the light purple of the clear morning sky gives a peaceful, eerie color to the silent, still snow. Mercury is fading in the gaining light.

We gathered up our weapons again for bivouac and also turned in our linen. When we turn in our sheets, we form a big queue and draw the white sheets over our shoulder. We resemble ghosts in the new snow.

As I was walking back from weapons pickup with Mike Snyder, he told me he was proud to be an American soldier. He says he feels good when we march in a company formation; he likes the pay he hopes the future will bring. He said back home in Ohio people told him that the army brainwashes people. I agreed. I agreed that my ideas of the army were wrong, very wrong. I probably would have joined years ago, but I was afraid. Afraid that I would be eaten up by this mindless evil, heinous juggernaut. This is the idea I had of what the army was. No humans in it at all. Just mindless evil beings controlled by some hidden supernatural force. What an incredible illusion.

It's 0800 hours. We were just practicing reporting to the officer outdoors and also an inspecting officer and also inspection arms. We are still on hold on whether to go to bivouac or not. I am writing on a smoke break. I reason that when the drill sergeant takes time to give the smokers time to do what they want to do, I should be able to do what I want to do. When the smokers hear it's a smoke break, they scramble wildly for their hats and coats and speed down three flights of stairs and light up. They barely have time to get outside, let alone smoke a cigarette. They come back more tense than relaxed, I have found.

Vieths is having ideas about getting out of the army. His wife wants him home. She is afraid of losing him. He wants to go to Monterey, but she wants him home in Minneapolis. He will be able to live with her in Monterey but thinks he will be scraping for money. His wife has a job in Minneapolis, but I don't think she wants to leave that or Minneapolis. People have been telling her how the service breaks up marriages, and that is the reason she says she wants him home. I suspect she just doesn't want to leave her job or Minneapolis.

6 February 1982

Did a little work this morning; we went to pick up weapons. Before our formation left, I went up to the barrack and retrieved my weapons card. I remembered my wallet, but I forgot to put my weapons card in it. Hensen saw me run into the barrack after it and asked me what I was going in for. (He was very pissed.)

"My weapons card, Sergeant Hensen."

"Well, why don't you put your name on one to two and two to three?"

I wasn't surprised to get such a reward for my genius, because yesterday I had forgotten my protective mask and last week I had lost my roadguard flashlight. Yes, I deserved it. I placed my name in the above slots, but Roberson's and Snyder's were there instead. So I got switched to ten to twelve. Much easier, because it is earlier. Many other trainees come up to me and ask me what I did to get two hours of fire guard, and I tell them. I am proud of my accomplishment, however, proud of the fact that *I* am the first to pull *two* hours of fire guard until further notice. It will give me time to get some writing and reading done, however, and some relaxing time to myself.

Hensen: "Do you get the idea that I got a hot date tonight, McCubbin?"

McCubbin: "Yes, Sergeant. Hensen."

Hensen: "You're right."

McCubbin said to us, "I almost lost my military bearing," and broke out laughing.

Hensen: "All right, Third Platoon, I'm glad to hear you all had a good time tonight. So I want you all to be ready for reinforcement training tomorrow. Good night."

Third Platoon: "Good night, Sergeant Hensen."

McCubbin: "Good night, Sergeant Hensen."

Hensen was smirking as he walked away. Barnes and Snyder came in laughing and slapped McCubbin's hand.

Vieths was called again just now, 7:30—it's an emergency this time. He says he is at his wits' end. What next?

Marshall and McKelvy are arguing and ready to fight, I think mostly because they have had too much to drink. It wouldn't be so bad, but they make so much noise arguing and if Sergeant Florez hears it, he is going to be pissed. It's impossible to calm down two megalomaniacs who have had a bit to drink. They each insist on having the last word, and as a result, there is a constant escalation of voices and hostilities. Neither is big enough to accept not having the last word.

Correction: just now Sergeant Florez came upstairs to the Third Platoon area and called all of us into the classroom. He quietly and sternly told us that he was the CQ tonight and that he hasn't had a day off for a long time and he is tired. He says he is not here to baby-sit a bunch of grown men. "Yeah, you've had a little bit of beer and you think you are hot shit! Well, I ain't scared a nobody. I ain't scared a Sergeant Hensen or

Sergeant Gibson or the company commander or the battalion or anybody. They can take my goddamned hat if they want to—they can take that son of a bitch. If anybody's arguing or fighting, bring it to me . . . I'll settle it."

At this point, I cut in and said, "I am glad you are, Sergeant Florez, because we sure the hell can't settle some of these people down—we just can't."

"I'll put it to you in plain English, folks. If I hear any more arguing or hear of any fighting in the Third Platoon area, I'm going to make y'all put on every piece of clothing in your locker and I'm going to take you out back and throw the whole PT book at you. I'm going to PT you until you drop—I don't give a shit. Just fuck with me tonight—just fuck with me." With that, he left our platoon area, and I heard him yell at the Second Platoon, who were making noise, "All right, I want everyone to get their long johns on, their boots, black gloves, field jacket, and pile cap. You jackasses are going outside!"

Hensen and the other drills decided to put us on pass today because he didn't think we had the energy and enthusiasm we needed. So the drills, and he decided to put us on pass—go out, have a few beers, have a good time! But don't go overboard. He asked how many were on pass, and everybody was (out of 59). He said this is the best platoon he ever had. When he was at Alpha 4-3, usually about 58 percent of a platoon would have their passes removed. Alpha 4-3 had 45 trainees. But this platoon is vastly different. He also said that out of 155 trainees at Alpha this year, 43 are not graduating with the company. He said he never looked forward to the end of cycle testing at 204 until this cycle. He said he is really looking forward to it. He said that those people at 204 don't have it out for trainees and don't go over there to see how many No Goes they can give. "I know most of them, friends that I have met, former drill sergeants. They want to see you pass; they want to see you do good. But then again they don't want to, and won't, pass an inferior trainee."

At this point I have a few days of notes in my red notebook, but it is lost.

11 February 1982

It's 10:05 P.M. On fire guard duty again. It looks as if the rest of the cycle I will be fire guard. I have had double duty of fire guard for the last four nights, not counting the night I had guard duty of a motor pool. Yes, it looks as if I am permanent here. I have done what nobody else has, double hour of fire guard for four nights in a row and it will probably be more. Other trainees are starting to say that I really have an MOS in fire guard. And that this is my AIT. Well, I do not mind it too bad since it gives me an extra two hours of time that I can get things done with, for instance, writing in my journal, washing clothes, reading some proverbs out of the Bible, writing letters, or just plain relaxing without having to go to sleep. There is a steel folding chair and a table. These are the only ones of their kind in our platoon area (besides the sergeant's office), and at this time we can sit on them. If you have ever gone a long time in an area where you either sit on a hard floor or lie on a bunk with dim lighting, you know how nice it can be just to sit on a steel folding chair with a backrest and write on a wobbly table. At this time I am going to jot down a few notes about what is on the wall. There are three eight-inch by eight-inch posters. One says:

> A sergeant is always right.
> He may be misinformed, inexact,
> Bullheaded, fickle, ignorant,
> Even abnormally stupid,
> But *NEVER WRONG.*

Below that is another larger poster, rectangular, about twelve by eighteen inches, colored red. On it is a cartoon picture of a gruff grizzly bear, gritting his teeth, pointing a thumb at himself, and he is dressed like a drill sergeant, complete with a Smokey the Bear hat. The poster reads:

> There are two ways to do anything around here—
> The *Wrong* way and *My* way.

Finally, below that one is another one—one of those make-believe forms with a large arrow pointing to a tiny box on the poster, and it says "Complaint Form." It could say "Compliment Form," I suppose, and be just as useful.

Today we had our end of cycle test and we are supposed to be soldiers now. Yes, soldiers. We, Charlie 4-3, today smashed the post records again for the highest number of MAXs on the test, plus-seventy, and highest percent of passing totally. We are the best—today we found out. Sergeant Gibson was really strutting today: "I told you all you were the best—and today y'all proved it."

Our reward was that we didn't have to say "one-two" now when we do the right-face, left face, about-face, or platoon halt. Some reward, I guess. Tomorrow we have our final DT test. Mama, mama, can't ya see? They made a soldier outta me.

Vieths did not go to the FOCT. He stayed back because he wants to get out. He really doesn't want to get out, but his wife wants him out. She is afraid the army is going to break up her marriage. And she is right—she will let it. She is acting childish and acting like she is cracking up. Vieths says he is at the end of his rope. He doesn't know what to do. I keep telling him, "We have no experience in these matters, but do what really feels right." The right thing, I believe, would be for him to go to Monterey and force his wife to also. If she doesn't go, then divorce her. She is forcing him to choose between the army and her when it doesn't have to be that way. She is thinking only of herself, I believe. She wants him to prove that he loves her by giving up all "this" for "her," and that is childish and stupid. A marriage cannot tolerate such imbecility. He could be living with her in about eight or nine days, but she won't listen to reason.

Two days—two nights ago, I guess—we had guard duty. Our company—Charlie—was assigned various guard-posts, and teams of two would walk their post. The shifts were two hours long. I served my guard duty with Guy Kawalski. We guarded a motor pool, and the guard route was at least one half-mile around the fence. The fence traversed glens and hills and forests. It was somewhat of a nature hike, since we saw a few rabbits. The moon was out bright that night. And cold, too. I found out that Kawalski was an astronomy aficionado like me, and we had much to talk about. But soon the talk turned to women, as usual, I guess. It does when around men.

12 February 1982

It's Friday, 2230 hours—I am still on my fire guard, the fifth night in a row of double-hour fire guard. I am beginning to think that Sergeant Hensen is being a little obnoxious about this whole thing. I miss out on at least two hours of sleep a night, and it is beginning to take its toll. I am getting bored with fire guard, bored.

Today we had our final test in our Basic Training careers. It was the final PT test. We were tested for the two-mile run, push-ups and sit-ups. I did not acquire superjock status, but I did my best. I did my best. I ran the two-mile in 12:23—I did forty-six push-ups and fifty-eight sit-ups. To get superjock, I would have needed ten more push-ups, and that's all. That is disappointing too, because from my first, second, and third PT tests I did not gain much in push-ups. My test scores were forty-two, forty-six, and forty-six. So, you see, I did not gain much there. I did gain in my sit-ups, thirty-three to fifty-eight. And I did gain on the two-mile 14:10 to 12:23, which *is a good improvement.*

After our PT test today (which, by the way, was in the brigade gym), we had chow, which deserves no compliment. Chow has been lousy for about the last two weeks. Ever since some companies have been graduating and leaving, the food has been lousy. We are the only company left in 4-3 (B+ Building) and as of next Thursday, we too will be gone. And we will be the ones in our dress greens at the mess hall eating and singing, "Na na na na, na na na na, hey, hey, good-bye. Na na na na, na na na na, hey, hey, good-bye." For the last two Thursdays we have watched companies graduating in front of us, and it made us envious and feel alone. First Alpha and Delta companies, and then Bravo and Echo companies. Each time we had to go through with watching them march away to graduation, it seemed so far away for us and something not really to get excited about. But in six days—six days—it will be our turn. But even those few days seem far away. I can pound the walls and say that and feel it. From where I sit right now, it seems far away. About all that is left for us now is cleaning our B+ gear and the barrack. We should get a big pass this weekend. I hope so.

Veiths got a big article today from Captain Liter about not going to the end of cycle testing yesterday.

Orfino is doing great, says he is glad he joined the army.

Today after chow, we had a class on the "Responsibilities of a Soldier." Captain Liter started us out with "Seats," and we screamed at a tumultuous level our company motto:

> "The difficult we do immediately.
> The impossible takes a little longer.
> Miracles by appointment only.

Duty, honor, country,
Drive on, Drill Sergeant, drive on."

Captain Liter was very happy with our performances at the EOCT and final PT test. So happy he said that he was speechless. He said the battalion commander was shaking his head, the brigade commander was shaking his head. They can't understand how C-4-3 does it so well. "The last five C-4-3s have broken all of the records," Captain Liter, grinning broadly, said simply. "Company, you have made me the happiest, proudest company commander on post—and especially thanks since this is my last cycle here." Captain Liter then turned the command for the remainder of the class to First Sergeant Williams.

First Sergeant Williams just started out by telling how proud he was of us, too, and that if he wasn't so old and ugly (he isn't), he would *cry!* Because we have done so well, he said the battalion commander called him into his office this morning. He wasn't sure why he was called in at first and thought he was in trouble. First sergeants usually do not get called in to see the commander—it just isn't done.

He said he got a perfect rating from the commander and it all was because of us—but also because of his cadre. First Sergeant Williams said that a film is usually shown here at this class, but he didn't want to use films because they just give you a chance to speak.

15 February 1982

Pulled two-hour fire guard last night. Some trainees have been staying out late, until 8:30.

Last two days we have been cleaning weapons—cleaned them until they were spotless. Finally had to give up my weapon (M-16) to the weapons room number 167—we had many good times together. I used to date a high-school queen; now I date an M-16.

Today we turn in all of our Basic Training gear—what a relief. But it is much work cleaning it.

Veiths was delivered a memo about receiving his Article 15 and a $100 fine. Hensen gave it to him and said wryly, "Keep it; read it; enjoy it!"

Yesterday Hensen was checking my weapon to see if it was clean, and he started giving me shit on how fast he could clean a weapon. "An hour and fifteen minutes, that is all it takes me." Big deal, an hour and fifteen minutes.

I said, "Sergeant Hensen, I ran out of cleaning equipment."

And Hensen laced back with, "Did I ask for a snivel, Stanley, did I?"

"No, Sergeant Hensen."

17 February 1982

It's 6:00 P.M. Last night before graduation here at Basic Training, and I am ready—finally—I am ready to leave, ready for a change and ready for California. I am really ready to git out of here. It is such a big relief I think that I could almost cry. No more of this fucking shit. No more miserable weather. No more drill sergeants. No more chow lines. No more road marches. No more noisy, echoing barrack. No more fire guard. No more mud crawls. No more gas chamber. No more gas mask. No more big mouths—none of the ignorant chip-on-the-shoulder types are following me to beautiful Monterey.

It has been a long cycle—I said that to Sergeant Hensen the other night, and he said, "I've noticed." I have tried my best—I really have tried my best, I feel—but I didn't shine head and shoulders above the rest. My mind is more spacy now than it used to be and my body is not in as good a shape as all of these young eighteen- and twenty-year-olds. I know that if I would have gone in when I was eighteen or so, my athletic tests would have been much higher. But I didn't do too bad. I would have liked to have been superjock but wasn't. But at least I made it to Track A.

The last few days we have had mostly foggy, dreary weather, and that has melted all of the snow. We are leaving Fort Wood as we first found it, dark and drab, but a bit more stereotypical—dank.

Tomorrow we graduate! We leave—I will be seeing some of my friends for the last time. Hawks, Snyder, Osterholm, Lamar, Veiths (he will probably be discharged). Those are some that are going to other stations for AIT. Many other of my friends will be following me to Monterey—Barnes, Callaway, Kawalski, Hansen, Jiminez and others (Colter, Bronas). I am sure I am going to enjoy California very much, and I am sure I am going to make many new friends. I am going to have to work hard, however.

Basic indeed was difficult. And now those same thoughts about the army are coming back—where will it lead?

Epilogue

I will give a short synopsis of what I thought of basic training. In short, it was good for me. I met a lot of nice people and got to experience firsthand what military life is like. It also built my confidence. It can be stressful, but they don't brainwash you. In my Basic Training I think it is significant that we didn't have one inspection. Our platoon sergeants said we were there to learn to become soldiers, not housemaids. Anybody else's Basic is bound to be different than that. I think the military in general puts too much emphasis on inspections. I am basing my feelings on stories I have heard rather than personal experience. Things should be clean but not taken to extreme like having inspections every day.

My life after Basic was not so successful. I went to language school. What a mistake! An officer at language school told us it was the hardest thing he has ever done. I didn't believe him. After three months of study I did. I had a migraine headache for two months, was seeing double, and was waking up at night shaking like a leaf. I tried everything to relieve the stress, exercise, aspirin, and psychotherapy. Nothing worked. Maybe I should have gone out on a week-long drunk. I dropped out of language school after three months and was medically discharged from the army a year later. In my travels since, I have met three other people who were in hospitals following language study. Almost everyone I talked to cracks up a little bit. But the VA still supports me, and I am thankful for that. Would I join again? Yes, in a heartbeat. But I would not recommend language school for anyone unless he has had extensive language training. Let me also say that if any of you are thinking about joining the army and have at least two years of college, then go in as an officer. You get a lot more respect. I was going to go to OCS after language school but never made it. Also, if you are thinking of joining, get down in writing everything you want out of the army. They will do it. If you want to go to Europe, then get it down. And get a folder, so you can keep all of your papers. You get a lot of them. I wish all of you soldiers and future soldiers the best.